SPARKS FLEW

SPARKS FLEW

WOSU'S CENTURY ON THE AIR

THOMAS M. RIELAND

TRILLIUM, AN IMPRINT OF
THE OHIO STATE UNIVERSITY PRESS
COLUMBUS

Trillium, an imprint of The Ohio State University Press.

Library of Congress Cataloging-in-Publication Data
Names: Rieland, Tom, author.
Title: Sparks flew : WOSU's century on the air / Thomas M. Rieland.
Description: Columbus : Trillium, an imprint of The Ohio State University Press, [2019] | Includes bibliographical references.
Identifiers: LCCN 2018055916 | ISBN 9780814213964 (cloth : alk. paper)
Subjects: LCSH: WOSU (Radio station : Columbus, Ohio)—History. | College radio stations—Ohio—Columbus—History.
Classification: LCC PN1991.67.C64 R54 2019 | DDC 384.54/53—dc23
LC record available at https://lccn.loc.gov/2018055916

Cover design by Mary Ann Smith
Text design by Juliet Williams
Type set in Adobe Sabon

♾ The paper used in this publication meets the minimum requirements of the American National Standard for Information Sciences—Permanence of Paper for Printed Library Materials. ANSI Z39.48–1992.

For the amazing ladies in my life: my wife,
Carol, and daughters, Kelly and Lindsay.
And for those who have chosen to dedicate
their careers to public media.

Thank you.

CONTENTS

PUBLIC BROADCAST NETWORKS AND NEW MEDIA (1960–2018)

ILLUSTRATIONS

PREFACE

I grew up in the idyllic community of La Crosse, Wisconsin, along the Mississippi River, the youngest of Gil and Angie's four sons. As a child, I remember one of my brothers, Allen, spending an exorbitant amount of time holed up in our basement with old radios and televisions. He somehow leveraged all that tinkering into a more than forty-year career in engineering and technology at Wisconsin Public Radio.

The Rieland boys may have been inspired by our dad's craving for some of the latest technology. We were the first in the neighborhood to buy a color TV in the early 1960s, and neighbors would show up to watch broadcasts of *Walt Disney's Wonderful World of Color.* Dad also had an 8mm film camera, a slide projection system, and a large Revere reel-to-reel recorder. I was probably thirteen years old when I pulled the Revere out of the attic to tape songs off the radio and then erase everything between the music and record my own pubescent voice mimicking a local disc jockey.

My radio bug reached its zenith while I was at the University of Wisconsin–La Crosse, where for one exhausting year I had announcing jobs at three area radio stations. The twelve-hour Saturday gig at WCOW in Sparta, Wisconsin, was most memorable. The station control room was upstairs in a rickety old house of this farm community, and my shift started at noon with the popular obituary report followed by a wacky music format that meandered from Glenn Miller to Johnny Cash to Olivia Newton John. I signed off at midnight, and spent the drive home through rural Wisconsin tuning to distant AM stations whose signals were skipping about the ionosphere, marking stations I heard on a notepad. Chicago station 89 WLS usually came in crystal clear, but 1090 KAAY in Little Rock and 1120 KMOX in St. Louis were less consistent. I would learn later that all were designated as clear channels, which were the most powerful AM stations in the country and protected from nighttime interference. One special night the signal skipped into my Ford Torino from San Francisco and then some Spanish station from who-knows-where. The year was 1976, and it was incredibly exciting to be twenty years old and a small part of this magical world of broadcasting.

*

In the fall of 2002 I was packing for the move to Columbus and WOSU after directing the Center for Public Television and Radio at the University of Alabama. As I was doing a final check of my office, I discovered a dusty, thick book wedged in the corner bookcase. I was surprised to find *Education on the Air,* a book published by The Ohio State University as part of the 1953 Institute for Education by Radio and Television. It was a fascinating read with reviews and transcripts of a dozen presentations from the institute. This was my first clue of the depth of broadcast lore buried in the Ohio State archives. I took the book with me to Columbus as a cherished treasure.

Through my years at WOSU and Ohio State, I'd learn that few stations or institutions would play as important a role in defining public service broadcasting in the United States.

Some believe the start of public broadcasting coincided with the 1967 Public Broadcasting Act. It was actually the tenacity of a few visionary educators nearly a half-century earlier that made the landmark legislation possible. This is dedicated to those pioneers.

The two most notable books about educational station history are focused on a single station—WHA at the University of Wisconsin. Both *9XM Talking* by Randall Davidson and *Wisconsin on the*

Air by Jack Mitchell are rich with detail and perspective. Those distinctive works were, in part, what inspired this book.

I thank the many alumni of WOSU who met with me to discuss their personal experiences. The professional staff at the Ohio State archives, especially Michelle Drobik and Kevlin Haire, helped me discover photographic and research gems and were always there to assist with my inane questions. Thanks to my development editor, Lynne Bonenberger, for her unwavering patience in helping create a presentable manuscript. I'm appreciative that Tony Sanfilippo expressed interest in this first-time author and thank all the OSU Press staff for their efforts to make this a reality in time for WOSU's century celebration.

One regret is the lack of space to mention the many dedicated staff, board members, and administrators who helped WOSU evolve into the substantive university and community resource it is today. I thank each and every one of you for your contribution, along with the many thousands in central Ohio who have financially supported WOSU through the years.

I've served two terms on the PBS Board of Directors and am constantly inspired by the passion and dedication of my station colleagues and the leadership of PBS president Paula Kerger and her team. Their efforts are matched by those within public radio stations and at NPR, the Corporation for Public Broadcasting, and our national advocacy group, America's Public Television Stations.

The ground was plowed and the seeds sown for public broadcasting a century ago in a few places like The Ohio State University. It's a story I'm privileged to tell.

Any profit from the sale of this book will go to support student experiences at WOSU Public Media to help train the next generation of public media professionals. Any errors of omission or commission are my sole responsibility.

Tom Rieland

WIRELESS TELEGRAPHY
(1895–1922)

CHAPTER 1

THE FIRST SPARKS

Robert Marriott is the wireless live wire.

—THOMAS EDISON

A roaring buzz of static echoed across the cavernous laboratory with irritating frequency, accompanied by bluish sparks. Some observers later compared it to the firing of a machine gun. To Bob Marriott, the effect was thrilling.

The young Marriott would sequester himself for long periods in an electrical engineering lab at The Ohio State University, working out the flaws in his patchwork spark gap transmitter. It was built to noisily send Morse code through the air to an equally crude receiver. Marriott was all of twenty years of age, but he was enamored with the idea of improving upon the recent invention of a young Italian whose magical machine would change the world.

Just a few years earlier, a shotgun blast across a quiet Italian valley had signaled a communications revolution. Guglielmo Marconi had been experimenting for years with sending electromagnetic waves through the air as a way to communicate. His discoveries were built on the work of Scottish physicist James Maxwell, who in the 1860s had predicted the existence of unseen electromagnetic waves, and on that of German scientist Heinrich Hertz, who engineered instruments to transmit and receive radio pulses. It was Hertz who invented the spark gap transmitter, but he would never realize the importance of his experiments to practical communications.

Marconi brought together the research of these and other scientists to formulate the ultimate experiment, based on thousands of tests using a combination of different coherers, receivers, sealed glass tubes, and metal filings. It played out near Bologna, Italy, in the summer of 1895. As Marconi sat on the window ledge of his lab in the attic of his parents' country villa, his older brother, Alfonso, and two workers walked from the villa across a valley carrying a wireless receiver device and an antenna. His brother also carried a shotgun.

Marconi had already proven it was possible to send coded signals using an old railroad telegraph key and transmitter to a receiver 100 meters away, the distance to the end of the villa's gardens. Experts at the time believed waves could be sent only in a straight line without obstructions, thinking that the main obstacle to sending wireless signals to a distant object was the curvature of the earth's surface. After his assistants walked over a hill to a distance of some 2.4 kilometers, or 1.5 miles, Marconi prepared his transmitter and pressed his telegraph key. The key interrupted the spark and sent three short blasts for the letter "S" across the valley. An instant later, Alfonso's shotgun blast signaled success.

It would take time for Marconi to prove to the world that true long-distance wireless communication, even across the Atlantic Ocean, was possible. But that day would shatter much of what was assumed about the use of radio waves for communications. Wireless was born.

ROBERT MARRIOTT OF RICHWOOD

Marriott Street in the Union County town of Richwood, Ohio, is a short stretch of pavement that dead-ends at a junkyard. The Marriott name was honored in Richwood because Henry and Rebecca Marriott helped establish the town. The couple also helped populate it, raising seven children who became teachers, builders, and farmers.

One son, Franklin, married a farm girl, Minerva Woodruff. In their fifty years of marriage, Franklin and Minerva would suffer through the deaths of two infant boys. Their third and last child, Robert, would survive and have a long and remarkable life.

Bob Marriott showed aptitude in the sciences and yearned to leave Richwood. In 1897 he was accepted to Ohio State University. His parents decided to join their cherished only child in the big city of Columbus. They found a thriving capital with more than ten thousand people, electrified streetcars, and the smell of hops emanating from the towering brick chimneys of the breweries south of town. It was a city of prominent structures, including the Columbus Lunatic Asylum, considered the largest building in the world under one roof; the massive Ohio Penitentiary on Spring Street; and the Statehouse, marking the center of the city at Broad and High streets. Among his many talents, Franklin Marriott was a skilled cabinetmaker, and he quickly found work at a furniture store. He later was involved in the founding of the Fifth Avenue Savings Bank.

A few miles north of downtown on the Ohio State campus, an armory and gymnasium designed by Columbus architect Frank Packard was under construction. A new president, William Oxley Thompson, would soon begin a long and distinguished tenure.

Bob Marriott joined a few hundred freshmen on campus and settled in. He had planned to pursue a career in physics and was quickly taken under the wing of the head of the department, Benjamin Thomas. In his first year Marriott approached Thomas with a proposal to focus his studies on the newly discovered X-ray waves, but his mentor had other ideas. Thomas captured Marriott's imagination by telling of the exploits of the young Italian Marconi, not much older than Marriott, who had just patented a wireless telegraph system. Marriott later learned that his professor may have had other motivations for steering him away from the mysterious X-rays. Thomas had partnered with Lewis Early, a Columbus physician who spent his career exploring the use of X-rays in medicine and surgery. The nature and properties of short-wavelength, high-frequency waves were unknown, but radiation exposure had proven dangerous. Thomas became one of its victims and was slow to recover from burns received during experimentation. Early continued to work with X-rays, and while it was said he cured many Columbus cancer patients, he would suffer a ghastly death from radiation exposure.

Marriott became the first and only student studying what was then called *Hertzian waves*. By 1900 he could often be found in front of a table of instruments in the engineering lab devoted to

FIG. 1.1. Robert Marriott's homemade receiver. Marriott was able to configure the receiver to print a dot when it received a coded signal.

electronic experiments. Coming on the heels of Marconi, Marriott's work was among the earliest documented student research. Learning how to send code into the air to be captured on an unwired machine in the distance became a sacred quest for Marriott. He was determined to find the key to an invisible world few understood.

Marriott built an impressive collection of wireless equipment to experiment with, including the principal instruments used by Marconi: an induction coil, mast, spark gap, relay, choke coil, tapper, and coherer. Marriott described the coherer in one of his papers as a 4-centimeter glass tube containing two electrodes spaced a small distance apart with loose metal filings in the space between. When the device detected a radio wave, the metal particles would cohere, or cling together, allowing direct current to flow through them. Since the metal filings in the coherer remained conductive after receiving the signal, the tube had to be "decohered" by tapping it with a clapper to break the filings apart. The receiver then would be ready to receive another wave. It was a sensitive outfit prone to failure.

Marriott had to plot a way to control his spark transmitter in the lab and adjust his receivers, which were placed at varying distances across the lab and beyond. As he described it: "The longer distance work about the campus was accomplished by fixing a clock pendulum so it would touch a pool of mercury closing a circuit and making the transmitter send a dot every second." When the spark gap was interrupted to send the dot, it resulted in a static squeal that echoed through the lab. The experiments logged by Marriott began with a distance of less than a foot between the transmitter and receiver. To decipher the characteristics of the radio waves, Marriott would change the length of the spark gap on the transmitter and the size and type of receiving antennas. He also would transmit at varying times of day and night. Spark gaps radiated a broad "dirty" signal that would later increase the problem of interference when other stations were set up in proximity.

The dots and dashes representing letters of the alphabet had been developed by Samuel Morse, a professor of arts and design at New York University. Morse proved that signals could be transmitted and received by wire over long distances, but his innovation wouldn't be widely appreciated by the public until the 1844 completion of the telegraph line between Washington, DC, and Baltimore.

The first long-distance phrase transmitted on the line was taken from the Bible: "What hath God wrought!" Three days later, the Democratic National Convention began in Baltimore, and Morse's operators sent the latest results back to Washington, where politicians and the press hovered around the telegraph station. Thus began a tremendous network of wired telegraph lines across America.

More than fifty years later, the idea that telegraphy could be used to communicate without the miles and miles of telegraph poles and wires shook the world. During his senior year at Ohio State, Marriott spotted a newspaper article about the new American Wireless Telephone and Telegraph Company in New York City and wrote to ask for a job. To his amazement, the president offered him a position if he could join the company immediately. Marriott packed up and left his parents and his studies for the trip east. Within a year, he had been promoted and assigned to Denver as chief engineer of the Pacific subsidiary of American Wireless.

As Marriott's success in wireless telegraphy grew, he became a point of pride back at Ohio State. Other students with an interest in electrical engineering would build their own reputations. Charles Kettering, a 1904 graduate, became one of the most accomplished inventors in the United States, with patents for automotive ignition and lighting systems. Kettering led the Delco Company in manufacturing those and other innovations and later became an Ohio State trustee.

In 1905 Lewis Warrington Chubb followed more closely in Marriott's wireless wake. An army brat, Chubb was born in Dakota Territory and spent his high school years in Cheyenne, Wyoming. How he and his brother found their way to Columbus is unknown, but he made the most of the opportunity. Chubb went on to do postgraduate work at Westinghouse Electric in Pittsburgh, and he never left. In 1919 his wife was killed when the driver of the car she was riding in was blinded by the headlights of an approaching vehicle. The tragedy prompted Chubb's groundbreaking research into polarization and the creation and patenting of filters on car headlamps to prevent glare.

THE CATALINA PROJECT

After Guglielmo Marconi successfully sent the Morse signal for "S" across the Atlantic Ocean in 1901, prospects for engineers in the new field seemed limitless. Bob Marriott quickly came to be regarded as one of a handful of wireless innovators in the US. He was the first to put into use the telephone and detector method of radio reception, a system that was a forerunner of the revolutionary Audion glass vacuum tube created by Lee de Forest.

Marriott's pioneering activity on Santa Catalina Island southwest of Los Angeles deserves special mention. In 1902 the kid who grew up in a small Ohio town received national attention for installing the first commercial wireless station on the Pacific Coast. The details of his experience reflect how early developments in wireless communications were viewed with both suspicion and wonder.

Catalina Island is 22 miles long and 8 miles wide. Marriott's station was the size of a small cabin and was perched 220 feet above the verandas of the Metropole Hotel, located in the town of Avalon near the southeast end of the island. His team had to blast off part of the hill to build the station house. Nearby was a 160-foot-high mast with a wire about a half-inch in diameter suspended at the top.

A similar wireless station was built on the pier at San Pedro on the mainland. Marriott gave the stations call signs: "A" for Avalon

FIG. 1.2. A middle-aged Robert Marriott, one of the early innovators of wireless telegraph and the founder of the first radio engineering society.

and "G" for San Pedro. Some speculate that this was the first time call letters were used for wireless.

For a young inventor who at Ohio State had struggled to send signals a few blocks, imagine the excitement of sending signals 26 miles over water. It was almost magical, even to Marriott, and it wasn't unusual to find doubters among the general public and the press. As Marriott put it: "The majority of the public that paid any attention to our efforts during the building and testing of the stations, seemed to be divided in its opinions of wireless experts. They seemed to think the wireless experts were supernatural, crazy, or crooked."

Marriott enjoyed telling of another incident that convinced many that wireless was not some parlor trick. Two men robbed the Metropole Hotel of cash and a case of champagne in the middle of the night then caught the 5 a.m. boat leaving Santa Catalina. The hotel owner located the island's wireless operator, who sent an urgent message to the mainland. As the boat docked at San Pedro, the thieves were arrested. Marriott described the publicized apprehension as a "spirited wireless comedy" for everybody but the robbers. It "shook any remaining disbelief out of the public," said Marriott.

ADVANCING WIRELESS AT OSU

After Marconi won the Nobel Peace Prize in Physics in 1909, many engineers, including Marriott, were troubled. Marriott's first employer had fought tooth and nail against the "foreign" Marconi Company and its allies, who seemed to be monopolizing the wireless industry. Marriott saw Marconi as simply a "sales engineer" out to convince everyone he was a profound scientist and an almost supernatural inventor. In truth, Marconi didn't have a strong theoretical background and worked mostly by intuition and experimentation. It didn't help that he also was known to be ruthless and distant, loyal only to his single-minded mission of building powerful transmitters. Marriott would never be generous in his attitude toward the Italian inventor who had inspired his career. He would, however, be very generous to Ohio State.

The first *Wireless Blue Book,* published in 1909, listed the known wireless stations throughout the country, including all ships and amateur stations. Nearly all the stations, including those in Cleveland and Toledo, were focused on providing communication with vessels.

The *Blue Book* gave Marriott an idea. He returned to Ohio State to encourage the Department of Electrical Engineering to take the lead on advancing research into using wireless to communicate between land-based stations across the country. In 1909, Marriott convinced his new employer, United Wireless, to contribute a complete new wireless system to Ohio State for research. The antenna for the station stretched above the newly built Robinson Laboratory from a point on the roof to the top of the chimney. For the first time the department offered a course in wireless telegraphy, and the first student club devoted to wireless was created.

Marriott is perhaps best known for a decision he made as a result of a scandal in his company. United Wireless became embroiled in fraudulent stock promotion schemes that eventually led to its demise. As an engineer, Marriott wasn't tainted by the scandal. Instead, he saw an opportunity and sent a letter to more than two hundred wireless engineers asking their opinion about forming an organization "to promote the authentic development of wireless technology." His efforts resulted in the creation of the Wireless Institute, the first radio engineering society, which published journals about research and test methods and equipment to expand the profession. The Wireless Institute later merged with the Society of Wireless Telegraph Engineers to create the Institute of Radio Engineers (IRE), with Marriott as its first president.

In 1963 the IRE and the American Institute of Electrical Engineers merged to form the Institute of Electrical and Electronics Engineers, which today is the largest technical support organization of its kind, with more 420,000 members in 160 countries. And it all started with Bob Marriott, the pioneer of the wireless world whom fellow Ohioan Thomas Edison dubbed "the wireless live wire."

There were only four graduates in the Ohio State electrical engineering department in 1901, and Bob Marriott wasn't among them, having left to pursue his wireless dreams a semester short of graduation. In 1914 the university waived his missing credits and bestowed on him a bachelor's degree in engineering. Marriott was living in New York City and couldn't attend commencement, so President William Oxley Thompson personally paid his five-dollar diploma fee. Upon notification, Marriott quickly sent Thompson a five-dollar bill in repayment. He also sent a letter mentioning an editorial in the *Ohio State Journal* claiming that Thompson was neglecting his duties while considering a run for Ohio governor. Marriott wrote that alumni in the New York area resented the editorial and were strong in their belief that Thompson was "the best president the University ever had."

Bob Marriott should be ranked among the most important Ohio-born inventors, but his name and accomplishments have been lost to time. Not only did he make wireless history; his efforts also created an institute that brought together the best and brightest in the field, advancing wireless into the age of radio and beyond. His contributions of expertise and technology to Ohio State helped the institution launch its radio curriculum and one of the earliest radio stations in the country.

Marriott lived in Brooklyn until his death at the age of seventy-two. He is buried with his wife Blanche and his ancestors in Claibourne Cemetery near his hometown of Richwood, Ohio.

THE *TITANIC*

In April 1912 the RMS *Titanic* sank in the North Atlantic after colliding with an iceberg. More than fifteen hundred people were lost in one of the deadliest maritime disasters in modern history.

The wireless operator on the *Titanic* sent distress signals via wireless Morse code, but a nearby ship's operator had retired for the night and the ship's captain ignored the flares sent up by the *Titanic*. The signals instead were received by a more distant ship, which came to the *Titanic*'s rescue, but too late for many.

In New York City, a twenty-one-year-old Marconi operator named David Sarnoff was at the telegraph key alerting any ships he could reach and informing the press and some relatives who crowded near his wireless station. As the only link to the disaster for the rest of the country, he instantly became famous.

The *Titanic* disaster changed Sarnoff's life, but it also changed the wireless field forever. As the *New York Herald* reported: "It appears that the disaster to the Titanic had no sooner been flashed over the seas than about every wireless instrument along the coast within range began sending and receiving with no thought of others, so the net result soon became a hopeless jumble, from which distorted and inaccurate messages were patched up in haphazard fashion and announced to the anxious world."

In a foreshadowing of similar issues with radio broadcasting in the 1920s, the "hopeless jumble" of wireless telegraphy demanded federal regulation. The US secretary of war led the charge to enact new laws requiring that a ship's wireless operation be constantly manned. To help clear the over-the-air clutter, the Radio Act of 1912 required the licensing for the first time of all wireless stations, including the experimental work in full operation at Ohio State.

CHAPTER 2

SOS COLUMBUS

The flood was second only to Noah's.

—MILTON WRIGHT, FATHER OF AVIATORS ORVILLE AND WILBUR, DESCRIBING THE 1913 FLOODING ACROSS OHIO

As the wireless act began to put new rules in place, Ohio State developed a powerful experimental wireless station with a transmission range of an astounding eight hundred miles. An experienced operator was hired to keep it operational. James Mercer had been with United Wireless before going on an expedition to the Amazon to build wireless stations in Brazil. He was back at Ohio State for further study and assisted with teaching a radio telegraph and telephone course. In February of 1913, with Mercer at the telegraph key, President William Oxley Thompson of Ohio State and President Harry Hutchins of the University of Michigan exchanged wireless pleasantries between Columbus and Ann Arbor, showing off the capabilities of wireless to the local press.

Within a month, such trivial uses of this new technology would be forgotten. In March of 1913, heavy rains with fierce winds and tornadoes caused severe flooding in fourteen states, including Ohio. Every major river east of the Mississippi flooded, and many bridges were washed away. Fallen telegraph and telephone lines isolated towns and cities in need of emergency assistance. People died from drowning, fires, freezing temperatures, accidents, and even suicide. Indiana, Ohio, and Pennsylvania were hardest hit. The event came to be known as the Great Flood of 1913.

In Columbus the downpour began on Easter Sunday. The cold, torrential storms intensified throughout the day, tapered off by midnight, and resumed in full force early on Monday. By Tuesday morning portions of downtown Columbus as well as Franklinton—the historic area to the west of the Scioto River known as the "Bottoms"—were about to be inundated. Conrade C. Hinds, former project manager in the Columbus Division of Water, chronicled what happened next: "Just before 11:00 a.m. on Tuesday morning, people living near the levee at the foot of Broad Street heard a peculiar roaring sound. The noise turned out to be the south bank levee that, after being weakened by the pressure of millions of gallons of violently rushing water, suddenly failed and gave way. A huge wall of water that resembled a small tsunami picked up speed as it gathered force after breaking through the levee and fanned out rapidly over the low-lying areas of Franklinton."

The waters were so powerful they destroyed all three main bridges in downtown Columbus. A streetcar was dislodged from its tracks and seven people drowned. Manhole covers were tossed in the air by geysers from the pressure on underground sewer pipes, and houses were lifted off their foundations and carried downriver. Some five hundred homes and buildings were destroyed under a surge of up to twenty feet. A police officer and others lost their lives trying to rescue people in the churning waters. Similar scenes played out in Dayton and communities across Ohio. The *Evening Dispatch* in Columbus printed a list of "missing and probably drowned," including numerous neighbors on Sullivant Avenue: "a mail carrier, a musician and Mr. and Mrs. George Eckert and their seven children."

The flooding isolated Franklinton from the rest of Columbus. Tom Green, a wire chief at American Telegraph, frantically attempted to route emergency calls along the few lines that were not down. Some calls went directly to the mansion of Governor James Cox, but most communication was hampered because of the miles of wires entangled in the rubble of the flood. Virtually all wired telegraph systems were down between Pittsburgh and Indianapolis due to the flooding across the region.

NEWS FROM THE HILLTOP

The Hilltop residential area just west of Franklinton became a refuge for those escaping to higher ground. It also was home to a

FIG. 2.1. Photograph of the Great Flood of 1913 in Columbus from the Baker Art Gallery collection.

brilliant fifteen-year-old wireless amateur named Herb Akerberg, who began using his homemade kit to send news of the impact of the flooding. The familiar dot-dot-dot, dash-dash-dash, dot-dot-dot international distress signal was part of many of his communiqués, which were sent repeatedly with the hope of finding a station to respond:

> "SOS Hilltop Business Men's Association wants city to send boats."
>
> "Supplies will last until about tomorrow."
>
> "People are suffering. Send this to Mayor Karb at once. SOS."

Veteran telegraph operators learn to identify each other by how they transmit the dots and dashes; almost like a fingerprint, no two operators use the key alike. At the Ohio State station, James Mercer instantly realized he was hearing an amateur well known to him by the initials H. A. The university's powerful transmitter began to relay Akerberg's messages about the destruction. "H. A. reported that a great many people on his street, Sullivant Avenue, were hanging in trees and in a very precarious condition," Mercer said. "These facts were immediately reported to the Ohio State Journal, who in turn promised to turn the matter over to the police." The *Journal* had one of the few wireless telegraphs in town.

Mercer also passed information to the *Journal* from an amateur wireless operator, J. B. Hyatt, in Mount Vernon, some fifty miles northeast of Columbus:

> "To: Governor Cox. Great flood: 100 lives lost, railroad lines all down; outside world cut off; . . . miles of track washed out."

The governor received the message via the *Journal* and sent supplies and provisions. Cox and his staff were doing business by torchlight and a single telephone line to Dayton.

Ohio State Signal Corps students soon set up a portable wireless station on a hill outside Dayton to report on the destruction in that city's downtown business area, including the flooding of Governor Cox's giant newspaper plant. The Signal Corps telegraphed details to the Ohio State station, which in turn relayed the information to the University of Michigan wireless, where national newspaper reporters were camped out. One return message offering help to Columbus came from Bob Marriott, who was now a radio inspector for the federal government in New York.

The Dayton floods claimed the lives of ninety-five people and more than fourteen hundred horses. Back on the Hilltop in Columbus, Akerberg stayed at his wireless for three days and nights. As C. B. Galbreath wrote in his *History of Ohio*: "He kept this constant vigil during heavy downpours of rain and at intervals waded in water up to his knees to the doorsteps of adjoining houses to get information to communicate to the city." Both Akerberg and Mercer were relieved from their work after some seventy hours at the telegraph keys. Operators from the Ohio National Guard and the Signal Corps arrived with five portable wireless sets from the War Department to begin broader communications throughout the region.

Mercer later said that his efforts had been hindered by continual interruptions from other amateur wireless operators: "We would tell UM [Michigan] to go ahead [with their code] and we would possibly be able to copy four or five words of a message, and then someone would break in, and it would be anywhere from fifteen minutes to three quarters of an hour before we could re-establish

FIG. 2.2. Postcards were created from images of the Great Flood in Columbus to send to friends and relatives.

communication." In a report issued after the flood, Mercer strongly encouraged Ohio State officials to support rigid enforcement of the laws regulating amateur wireless stations.

On campus, an experimental poultry farm and hog barn on the west side of the Olentangy River had been seriously flooded, and the head of the poultry department, F. S. Jacoby, had barely escaped the rising water. Otherwise, there was only minor damage to campus. Faculty, staff, and students focused instead on coming to the aid of downtown and Franklinton residents, with some students attempting to rescue stranded residents and narrowly escaping tragedy.

More than one thousand students—a quarter of the university's student body—went to work in relief missions, making food and other provisions for the newly homeless and passing out supplies from a school on the city's west side. Men from the university's cadet battalion volunteered for guard duty on the edge of the stricken area. The chair of the Columbus Relief Commission, Samuel Prescott Bush, said he "could not begin to tell how we have appreciated the work of the University." Bush was president of Buckeye Steel Castings, a prominent employer on the south side of town (and, incidentally, the grandfather and great-grandfather of two future US presidents).

HEROES AT THE KEY

The Great Flood of 1913 in Columbus, Dayton, Delaware, Zanesville, and other towns across the region remains the most devastating weather disaster in state history. At least four hundred sixty people died in Ohio, and some forty thousand homes were destroyed. As the waters receded, a photographer from the Baker Art Gallery took pictures that later filled a booklet called "Flood Views 1913." The twenty-eight stark images were accompanied by captions: "Rescue Boat on Broad Street," "One of the Many Victims," and "Pool where Officer Sexton was Found." Within a few weeks, postcards with photos of the catastrophe were being marketed.

The stories of Mercer and Akerberg and their nonstop hours "at the key" became part of wireless lore. The operator at the University of Michigan wrote the following for *Modern Electronics* magazine: "Mr. J. A. Mercer, the able operator at the Ohio State University, . . . stuck to the job for 70 hours, when he collapsed and was temporarily relieved by some of the signal corps operators. As a class I wish to praise the amateurs. They have shown the world that wireless can be of the greatest service when called upon."

The flood marked one of the first local uses of emergency wireless "radio" in the US, with young amateur telegraphers thrust into the limelight. The experience helped propel amateurs to create an informal system of relaying messages in times of emergency. The American Radio Relay League, founded in 1914, played

HERB AKERBERG

Herbert V. Akerberg, the young wireless operator from the Hilltop who transmitted news of the 1913 flood, graduated from West High School in Columbus, served in the Signal Corps during World War I, and returned to the city to build commercial radio station WAIU, one of the original CBS radio affiliates. The station today is called WTVN.

Akerberg was an impressive engineer, creating a mobile car radio as well as an audio mixing panel for adjusting multiple studio microphones while recording performances. He moved from engineering into management and became a shrewd businessman and a fixture at Black Rock, the CBS headquarters in New York City, where he handled the commercial radio network. Akerberg later served at the right hand of CBS Television president William S. Paley as vice president of affiliate relations.

an important role in reporting about the Mississippi River flood in 1927 and the Ohio River Valley flood ten years later. The league is still going strong today, with thousands of amateur radio members across the country.

Ohio State's wireless coverage of the flood of 1913 was recognized time and again as an example of how the university could use its research and technology prowess to assist the region. Its efforts were said to lay the foundation for its "policy of public service" and for the later development of radio broadcasting.

EXPERIMENTAL 8YO

It wasn't a coincidence that a teenage boy in Columbus was reporting on a flooding disaster from his small, homemade wireless and antenna kit. Wireless telegraphy enticed very young radio enthusiasts who were attracted to the technology and the low cost of entry to participate. *Amateur Works* magazine included information on how to build a simple transmitter and receiver. By 1912 there were at least one hundred twenty radio clubs in the US. Many amateurs operated with skill and efficiency in those early days, but as the hobby expanded, the growing number of operators became a nuisance to commercial stations and fellow amateurs. Some amateur stations assigned themselves their own call letters and transmitted however and whenever they chose, interfering with serious operators.

Under the Radio Act of 1912, passed after the *Titanic* disaster, wireless stations were required to have government-issued licenses to operate. That wasn't a major issue for most operators because the process was simple, much like applying for an automobile registration. However, amateur stations were assigned wavelengths inferior to those reserved for government, commercial, and educational use. Amateur Morse code stations flourished across Ohio, with stations licensed in homes in Springfield, Cincinnati, Hamilton, Mount Vernon, and Canal Dover, along with high schools in Cincinnati and Lima and the Goodyear Tire and Rubber Company in Warren.

The Radio Act designated certain frequencies for educational institutions such as Ohio State to use for experiments to develop the science of radio. The work was assigned the uncluttered frequencies of 500 to 1500 kHz AM and given government-issued call signs.

Before any call signs were attached to Ohio State, the university announced new advancements in wireless research, including once again upgrading the transmitter and receivers on campus. Engineering faculty led by Roy Brown installed an aerial in 1914 that caught the attention of the student newspaper: "Have you noticed the large new aerial for the wireless station in Robinson Laboratory, which has just been put up on the new 200-foot smokestack? The highest point is about 185 feet high and the lowest is 80 feet high. It is the second largest aerial in the state. Only one is larger, and that is the Government station at Cleveland which is used with the Great Lakes boats. Messages within a radius of 2500 miles can be received."

Continuing a long tradition of "Electric Shows" sponsored by engineering faculty and students, a three-day festival in 1914 featured one hundred exhibits at Robinson Lab. The program included demonstrations of cooking eggs on ice, the latest in X-ray machines, and electric light bulbs not yet on the market, along with a wireless demonstration of how to control and sink model warships, all for just 25 cents.

The experimental station at Ohio State was already well established when it was officially licensed in March 1915 and assigned its first call sign: 8YO. The station was sending regular coded transmissions by students and faculty every Wednesday and Friday evening. It became obvious to the station's engineers that radio waves reverberated for much longer distances during the nighttime compared to daytime. Ohio State exchanged messages with stations from afar at night, but its strongest collaboration continued to be with wireless colleagues at the University of Michigan. Imagine the cutthroat drama of the wireless Morse code chess contests between the two rivals.

Midwestern-based universities with land-grant service traditions viewed wireless as a way to extend their mission to disseminate important information and knowledge to the general public. When Congress passed the Smith-Lever Act in 1914, the national Cooperative Extension Service came into existence at all land-grant universities. Extension agents spread across much of rural America to help educate and inform farmers about advances in agricultural practices, home economics, and the 4-H youth program. Pioneer educational stations immediately developed relationships with the new Extension Service units. What better way to provide services to the rural parts of a state than by sending information from a single source such as a wireless telegraph?

In 1916 the US Weather Bureau decided to develop a wireless network and joined with four powerful stations to distribute forecasts to farmers across the Midwest: 8YO at Ohio State, 9XM at the University of Wisconsin, 9XN at the University of North Dakota, and the Great Lakes naval installation in Chicago. These organized efforts helped establish both the public service model of universities using their technology to extend their mission and the modern broadcast model of sending one message to be received by many.

Ohio State's 8YO began tapping out coded weather forecasts and farm-market reports from 11 a.m. to 1 p.m. daily to wireless receivers operated by farmers trained in code reception and desperate for such information. The code was also received in county extension offices and circulated throughout farm communities. Connecting rural Ohio in partnership with Ohio State Extension would remain a significant programming strategy for decades to come.

MUSIC IN THE AIR

Wireless code work was always seen as a transition to the ultimate goal of sending voices and music through the air. The spark gap transmitter went through multiple stages of development, but experiments in carrying voices were largely unsuccessful because voice transmission required continuous waves.

Most historians agree that one of the earliest transmissions was Reginald Fessenden's broadcast on Christmas Eve in 1906. Fessenden had been a chemist in Thomas Edison's lab before becoming a professor of electrical engineering. His audience was likely shipboard Marconi wireless operators, who had been put on notice that something special was going to come across their receivers. Fessenden's program started with a short speech, followed by a phonograph playing "Handel's Largo" and a violin solo with Fessenden himself playing "O Holy Night." The choices foretold 1920s radio programming with the voiced advance promotion, the choice of "quality" classical music from one of the masters, and a live instrumental performance.

By 1916 inventor Lee de Forest had established a regular schedule of evening concerts of phonograph music, though reception quality was extremely poor. It was de Forest who helped propel radio's viability after his creation of an electrical amplifying vacuum tube. But while experiments with regular broadcasts of voice and music were reaching some maturity, something else was in the air that would put a halt to everything: the Great War.

CHAPTER 3

WWI AND COLUMBUS RADIO

> This is a war of applied science, not of physical force; a war in which intelligence, inventive genius and moral force will triumph in the end.
>
> —WILLIAM OXLEY THOMPSON, OHIO STATE PRESIDENT

On April 9, 1917, Ohio State received a letter from the US radio inspector ordering the dismantlement of all its wireless equipment. With the nation's entry into World War I, President Woodrow Wilson had issued an executive order shutting down every wireless station for security reasons, as wireless telegraphy could be heard from quite a distance. Just months earlier, the 8YO experimental station had been one of a chain of powerful wireless relay stations developed as part of the American Radio Relay League's plan to serve the country.

Ohio State appealed by telegraph, asking permission to retain the wireless for government use in an emergency, but the request was denied. Station 8YO was out of business, and the equipment would become part of the army's effort to teach communications. The government also took control of railroads, developed a war conscription program that affected more Americans than at any other time, and cranked up a propaganda machine to promote the sale of war bonds and to demonize Germans. Direct impacts in Columbus were the closing of the German schools south of the city and the shutting down of the last German-language newspaper.

By June 1917 the Ohio State campus was awash with cadets and civilian instructors. One of six university flying schools in the country had been established within the new School of Military Aero-

FIG. 3.1. The Ohio State aviation building, completed in 1918, became home to the electrical engineering communications lab in 1920. It housed the offices and studios of WOSU radio until 1970.

nautics. The squadrons of cadets underwent eight weeks of training in photography, signaling and radio, gunnery, airplane engines, aerial observation, and piloting. The school also provided the only training site for officers of the Army Balloon Corps. During the war, tethered balloons with Army scouts in wicker baskets were floated a half mile into the sky to spy on the enemy.

The radio branch of the Signal Corps unit was established, directed by the fittingly named James Code. Given Ohio State's leadership in wireless telegraphy, the army focused on courses in radio transmission and reception. The growing School of Military Aeronautics shared cramped quarters with other Ohio State units until a new building was completed in 1918.

The new aviation building was designed with an unusual sawtooth skylight with north-facing windows placed vertically, allowing for maximum indirect natural light. As developed by university architect Joseph Bradford, the idea was to minimize shadows as students worked together on airplanes. With 15,000 square feet of open space, the lab's hangar was large enough to house four airplanes and two rooms designated for the study of wireless. Cadets regularly took planes apart and reassembled them as part of their classes. The planes periodically would be rolled down the hill from the aviation building to a grassy field just east of the Olentangy River for flight tests and training. The airstrip wouldn't survive beyond the war because the university found a much more lucrative use for the field. It became the site of the massive Ohio Stadium.

As the war began, Ohio State president William Oxley Thompson spoke of the need for students in engineering to apply for the special military course in radio. He encouraged them to be ready when the call came from the armed services and to feel honored to be called. More than four hundred students decided not to wait for government action and withdrew from Ohio State to volunteer for service.

INTRODUCING ROBERT HIGGY

To help lead instruction in wireless communications in the new school, the army tapped an unlikely youngster, a local high school student named Bob Higgy. As a ten-year-old living in Lima, Ohio, Higgy received his wireless operator's license and used a homemade spark coil transmitter to send code across the state. His father was a salesman and moved the family to Arizona and California. Along the way, young Higgy built a reputation with operators across the western US for his technical prowess. He was regarded as a "smooth code operator" for his deft touch on the key and his ability to send code quickly using the many telegraphy abbreviations of the time. When Higgy's family moved to Columbus, the Signal Corps recruited him to teach students the science and art of wireless telegraphy. He was hired as part of Ohio State's clerical staff but was put to work as a part-time instructor, paid at a grade equal to that of a first lieutenant.

"ON WISCONSIN"

The war may have disrupted wireless progress at Ohio State, but it seemed to have little impact on one of its chief rivals in the field.

At the University of Wisconsin, engineers had recognized the limitations of the spark gap transmitter and had begun experimenting with three-element vacuum tubes invented by Lee de Forest for both transmission and reception. Because the tubes were not commercially available, Wisconsin engineers, led by Earle Terry, designed and built their own during the war years. They worked on the assembly of the electronic components and even hired a glassblower to create the tubes. The method sometimes took twenty hours, and a tube often would fail at the end of the process when a plate inside melted, a filament burned out, or the tube cracked.

Terry was able to resume Wisconsin's radio work during the war through a special arrangement with the navy that included experiments with testing reception of wireless code by submerged submarines in the Great Lakes. In 1917 the school's experimental 9XM station began perhaps the earliest university-based broadcasts of voice and music with Terry's tube transmitter.

VOICES AND MUSIC FROM CAMPUS

At Ohio State, a bright sign of radio progress came soon after the completion of the new aviation school in 1918. Roy Brown and R. C. Schott of the electrical engineering department worked with the army radio class to develop transmission antennas that could be used near frontline trenches, since stringing high wires during battle was unfeasible. Wires were stretched between three poles that were six feet high, and voice transmissions were sent across the Olentangy River to receiving equipment in a barn near what is today the Fawcett Center. It was the first documented transmission of voice communications on campus. The invention would never be used. Within six months, Germany surrendered and the war was over.

New federal radio assignments after the war gave Ohio State the call letters 8XI for its experimental radio operation. The designation placed it among the few universities providing serious instruction and research into radio communications.

In his first year as a student at Ohio State, Bob Higgy joined Roy Brown in the electrical engineering lab to build a vacuum tube transmitter capable of reliably sending voice and music into the air. As an experiment, they started dropping steel needles on 78 rpm phonographic records. It was the start of Prohibition and the birth of the Jazz Age, and Higgy played his favorite jazz records every Wednesday night to an audience of some two hundred amateur wireless receivers.

Although these were the first regular over-the-air broadcasts of music in central Ohio, most people didn't understand the impact. The Ohio State *Lantern* didn't know what to make of the new technology, so in an article dated March 8, 1920, it branded Higgy "the first wireless telephone maker in Columbus." Later that year operators at 8XI began voice broadcasts of market and weather reports intended for farmers.

At the annual Electric Show in 1920, students and faculty were thinking especially big. The show spanned six days, with special invitations to Ohio inventors, including Thomas Edison. (He didn't attend.) It was held for the first time at Memorial Hall on Broad Street, attracting a curious downtown crowd. The show provided "wireless signals from all parts of the world amplified for visitors, including reception of jazz played live in New York and Cleveland," according to the *Lantern*. An antenna was erected between the flagpoles of Memorial Hall and the nearby Elks Club to bring in the distant radio signals. Some of the entertainment came from closer to home, with music from the Ohio State Glee Club transmitted from 8XI on campus.

It would take some time, but the verbiage "wireless telephone," "wireless telephony," and "radio telephone" would soon be swept away for a more modern and simple word: *radio*. The word may have stemmed in part from the telegrams sent by wireless before the war, which were called *radiograms*, or *radios* for short. It may also have come from the "radiating" of electromagnetic waves from a transmitter.

Engineers at educational institutions in the Midwest can lay claim to the common use of the term *broadcasting*. The word echoed the agricultural term for spreading seed across a large field and was first used prior to the war to describe the transmission of Morse code from one transmitter to many receivers.

This newfangled technology, "radio broadcasting," was catching fire across the country and on the Ohio State campus. In 1920, the first of many regular appropriations was granted to the College of Engineering to support radio communications. President Thompson approved the hiring of the first radio operator, Paul Breeze, who shared the position with his brother John. They manned station 8XI for 75 cents a night.

KDKA

As two Ohioans fought for the US presidency in the fall of 1920, radio coverage popularized the new notion of broadcasting to a listening public across a geographic region. The center of the radio universe on November 2 at 8 p.m. was a makeshift shack atop one of the Westinghouse manufacturing buildings in Pittsburgh that included a small transmitter, turntable, and microphone. Broadcast of the election results involving two wealthy newspaper publishers—Senator Warren Harding and Governor James Cox—caught the attention of the country.

Westinghouse had received a special license from the Department of Commerce days prior to the election to start its service with the call letters KDKA. It was a transitional moment. As Erik Barnouw wrote in his seminal work on radio history: "It was the first time radio was an activity for everyone, not just the technical prodigies." The plan to broadcast the 1920 election returns was hatched as a public service but, more importantly, to sell as many Westinghouse home radio receivers as possible.

Behind the scenes, the newly appointed manager of the Westinghouse radio department was Ohio State graduate Lewis Chubb. On November 2 Chubb was busy installing receiving equipment at the auditorium of the Edgewood Club outside Pittsburgh, where a crowd gathered for their first experience of listening to a real radio station. Chubb called in to KDKA on a regular basis to report on the quality of the signal and the broadcasts.

The potential of radio had begun to reveal itself. The KDKA broadcast was heard not just in the Pittsburgh area but by people across the country. All a listener needed was a cheap receiver, an external antenna, and some talent for tuning. KDKA's signal, without any competition for the airwaves, boomed into Columbus and across Ohio, where interest in the election was especially high. The *Cleveland Plain Dealer* reported that more than six hundred radio amateurs held election tune-in parties at their homes. It was the first presidential election in which women had the right to vote in all forty-eight states, and millions of them turned out. Harding won a landslide victory.

KDKA made further history in 1921 by broadcasting the first live-voice play-by-play of a football game, a matchup between the University of Pittsburgh and West Virginia University at Pittsburgh's Forbes Field.

The KDKA accomplishment received substantial attention in Columbus. When it comes to football, even in the early 1920's, Ohio State refused to be second to none. Soon the Department of Electrical Engineering received an exorbitant special state appropriation of $40,000 (more than $500,000 if valued today) to develop an improved radio station laboratory to, among other things, broadcast OSU football to alumni. The funds also allowed the department to build soundproof studios in what would now be called the Communications Laboratory within the space deserted by the defunct WWI Aeronautics School. The electrical engineering department soon filled the rest of the space with classrooms, labs, and a machine shop.

After Bob Higgy helped build Ohio State's experimental station in 1920, he decided to develop his own station. He founded the first local amateur radio club and assembled a transmitter to broadcast live concerts of student musicians from his parents' home on East Frambes Avenue near campus. As the *Lantern* reported of one concert: "Nine musical selections were played, many by a seven-piece orchestra" in the Higgy living room.

In December 1921 Higgy was enticed by a new opportunity well beyond Columbus. He decided to interrupt his college education and move to Hartford, Connecticut, to become assistant editor of the amateur radio magazine *QST*. The magazine had been reporting on Higgy's wireless prowess for years and offered him an opportunity to spread the word about the potential of amateur radio.

WEAO ON THE AIR

As Ohio State's winter session began in 1922, radio courses were overrun with students. New commercial stations were under construction in Columbus and across the country. Radio engineering,

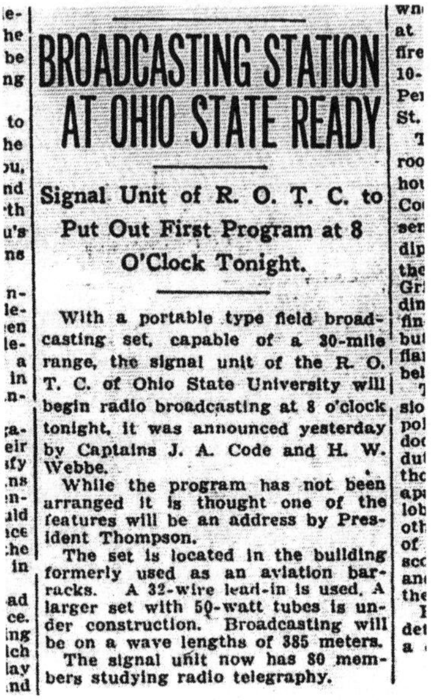

BROADCASTING STATION AT OHIO STATE READY

Signal Unit of R. O. T. C. to Put Out First Program at 8 O'Clock Tonight.

With a portable type field broadcasting set, capable of a 30-mile range, the signal unit of the R. O. T. C. of Ohio State University will begin radio broadcasting at 8 o'clock tonight, it was announced yesterday by Captains J. A. Code and H. W. Webbe.

While the program has not been arranged it is thought one of the features will be an address by President Thompson.

The set is located in the building formerly used as an aviation barracks. A 32-wire lead-in is used. A larger set with 50-watt tubes is under construction. Broadcasting will be on a wave lengths of 385 meters.

The signal unit now has 80 members studying radio telegraphy.

FIG. 3.2. The *Columbus Evening Dispatch,* April 24, 1922, announcement of the on-air debut of regular programming on the Ohio State radio station.

operations, and announcing were becoming popular professions. More than one hundred forty students took courses related to the construction and operation of stations and radio law to prepare themselves for federal exams to receive amateur licenses.

On April 24, 1922, at 8 p.m., Ohio State staged a special radio event, including a grand announcement by President Thompson, which was broadcast using the Army Signal Corps mobile transmitter on campus. (The transmitter in the electrical engineering department was under repair.) The evening kicked off regular programming using the 8XI experimental call letters and included musical selections played on 78 rpm records courtesy of the Victor Talking Machine Company. Selections included the Paul Whiteman Orchestra playing "Poet and Peasant," "Honolulu Eyes," and "Ty-Tee." A newscast with information gathered by the *Columbus Citizen* newspaper included a report of a fire at a downtown shirt factory. Coverage was estimated at a 30-mile range.

By all historical accounts, the broadcast was successful in that Ohio State could lay claim to being the first radio station in Columbus, even though it had yet to receive its official federal license or its call letters. Such a claim was certainly justified, given the early history of voice and music broadcasting in 1920 and 1921, but the

FIG. 3.3. President William Oxley Thompson at the microphone of Ohio State's new radio station, April 24, 1922. It was still an experimental station at this point and would gain its first call letters, WEAO, in June.

timing of the press event was curious. Why not wait until the official call letters were assigned? Perhaps the answer was that Ohio State's license filing had somehow been delayed, and two commercial radio stations in Columbus were days from receiving notification of their call letters.

Whatever the reason, Thompson dramatically announced: "We are starting tonight the first of a series of programs of entertainment and instruction for the citizens of central Ohio. Happily, Columbus' first radio program is being broadcast from The Ohio State University."

Five days later, the first federal radio license granted in Columbus went to commercial radio station WBAV, which soon changed its call letters to WAIU to reflect its parent company, the American Insurance Union. WAIU was a charter member of the CBS Radio Network, being one of the sixteen stations that aired the first CBS network program on September 18, 1927. The station today broadcasts as WTVN-AM.

Two weeks later, WCAH became the second licensed station in the city. The *Columbus Dispatch* bought the station in 1929 and changed the call letters to WBNS, which stood for Banks, News, and Shoes, the types of businesses controlled by the locally prominent Wolfe family.

The Ohio State station was finally licensed on June 3, 1922, with the initial call letters assigned as WEAO. It was officially the sixth educational radio station in the country, although it can justifiably claim to be one of the first. Jack Mitchell—public radio pioneer, historian, and former director of Wisconsin's educational station—had it right when he said: "In reality, no broadcast pioneer can identify precisely when a station began. Each station developed incrementally starting with point-to-point messages in Morse code, eventually adding voice and music."

The first five educational radio licenses went to the universities of Wisconsin, Minnesota, Illinois, Purdue, and Iowa State. It was a stark indication that the birth of educational radio was the province of major universities that tapped brilliant, ambitious faculty and students within physics and electrical engineering departments to give birth to the new medium.

EDUCATIONAL RADIO AND TV (1922–1960)

CHAPTER 4

A RADIO EXPLOSION

> The air is throbbing with the songs of famous soloists, with instrumental music from orchestras, with lectures, sermons, news bulletins, market reports, weather forecasts, and political speeches. The ether has become one vast auditorium. And all that is necessary to make the silence speak is to have one of the magic radio receiving sets that are the sensation of the age.
>
> —*MOUNT VERNON DEMOCRATIC BANNER*, 1922

Flowery prose in the local newspaper greeted the arrival of radio in Mount Vernon, Ohio, in 1922. Ohio State physics professor Frederic Blake was in the Knox County town as part of the Redpath Chautauqua circuit to showcase the magic of radio waves through demonstrations to overflow audiences in communities throughout the state. Blake's setup included a tent filled with radio receivers with aerials set to receive broadcasts from stations across the country, including his own university station, WEAO.

Radio's growth was phenomenal. On a single day in 1922 the number of radio stations in the US exploded from 28 to 570. Anyone with some means could start a station. The Nushawg Poultry Farm in New Lebanon, Ohio, developed a station to broadcast information on hog farming, dairy feeding, and poultry culture. "Within the first three months we had over 4,000 inquiries about our products," bragged W. N. Nushawg, poultry farmer and radio station entrepreneur.

With early radio there was always the chance of poor reception stemming from interference or issues with receivers, antennas, or weather. At an Ohio State carnival in the spring of 1922, one of the engineering fraternities posted a sign outside its tent advertising the opportunity to put on headphones and hear the famous and powerful KDKA station from Pittsburgh. The tent soon was full of townspeople who paid a 10-cent entry fee for an experience with the new medium. When the operator tried to tune in KDKA, however, all he found was static. The frat boys were in a panic but quickly engineered a plan. While some students talked to the crowd, others tracked down the radio operator in charge of Ohio State's transmitter. Fifteen members of the marching band were recruited to sit before a microphone in a tiny room and play. An engineering student became the announcer for the day, describing the music as coming "live from KDKA in Pittsburgh." No one was the wiser.

As euphoria about this new wonder spread, radio sets were quickly coming on the market: the Aeriola, the Radiola, and many other crystal and tube receivers. In February 1922 President Warren G. Harding introduced the first radio to the White House with fanfare. Later that month, Commerce Secretary Herbert Hoover, labeled by the press as "supreme chief of the ether cops," convened a small conference of industry and governmental officials on the rapid growth of radio. Hoover emphasized that public interest in the new medium was paramount and that the vital decision of who should be allowed to broadcast for what purpose needed to be addressed. By the end of the year, there were more than six hundred licensed stations, and massive chaos in the air reigned as stations fought for frequencies. Hoover called a series of conferences in the months to come to address the many issues resulting from the boom.

When the Ohio State station was licensed in June 1922, the federal government randomly assigned the call letters WEAO and ordered the station to broadcast at 360 wavelengths or 833 kHz on the dial, the only local frequency allotted for music transmission. All three Columbus stations at the time had to agree to share that one frequency by scheduling their programming at different times of the day and night. Because WEAO wanted to provide a news and information service, it was required to change its frequency when it broadcast market and weather reports. Once the reports were completed, WEAO switched back to the "music frequency."

Those owning a radio receiver would tune in to a jungle of voices and music, one overpowering the other depending on where the listener was located. One major city had twenty-five stations, all sharing time on one AM channel. Some of the large commercial

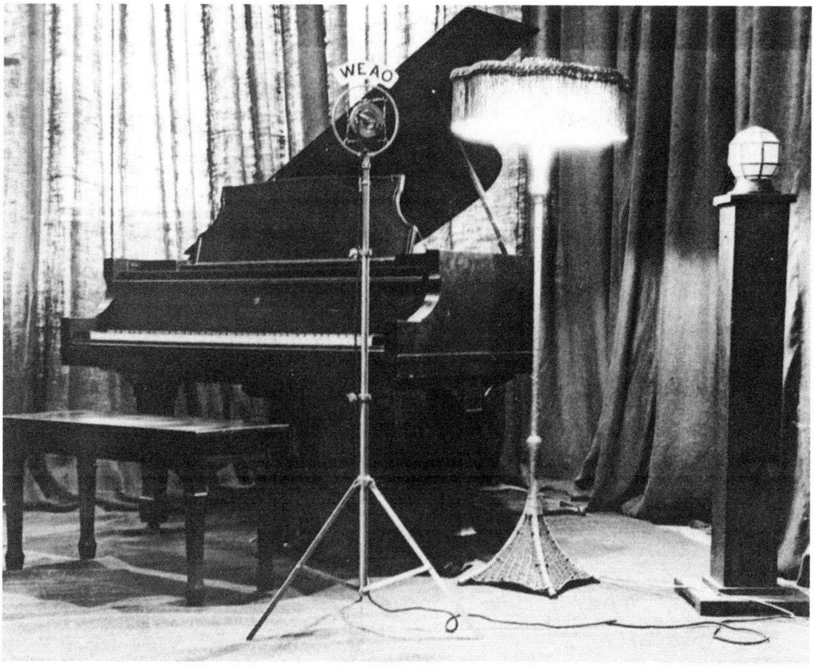

FIG. 4.1. A promotional brochure written by Ann Charles and Robert Higgy described the WEAO studio as "arranged with draperies and hangings to give the proper acoustical effects, so necessary for the proper transmission of any sort of entertainment, a grand piano and suitable furnishings."

stations ignored time-sharing agreements and blasted through with giant, 50,000-watt transmitters, wiping out any competition.

As complaints from listeners and stations mounted, the Department of Commerce finally opened up a swath of frequencies from 550 to 1350 kHz and began to reassign channels. The 100-watt outfit at WEAO had coverage of about one hundred miles from the Columbus city limits in the daytime, but it magically reached homes across the Midwest and beyond during evening hours as the signal hopped about the night skies. The farm market and weather reports remained the bread and butter of the new station. Ohio farmers enthusiastically snapped up radio receivers. Eleanor Farjeon, a poet of the time, captured the importance of these radio broadcasts to rural communities:

And now the farmer knows what feast
Of sun is coming from the East,

FIG. 4.2. Announcer Robert Coleman in front of a Western Electric carbon microphone at WEAO. Coleman later led the educational station at Michigan State University.

Knows when his furrows will be blessed
With feeding rainfalls from the West.

He from the Forecast now will know
When broadcast he his seeds may sow,
When it is wise to cut his hay,
And when to cart the corn away.

The dean of engineering at Ohio State appointed Charles Wright to direct the station part-time, all the while keeping his teaching load in the department. Wright described the potential of WEAO to a *Lantern* reporter in August 1922: "While the novelty of radio is responsible for its appeal to many people, there is a valuable and permanent service which it can supply. The station will be one of the best means of service of the Ohio State University to the people of Ohio." Wright would prove an enthusiastic and effective leader in WEAO's earliest years, advocating strongly to an administration that still seemed unconvinced of radio's value and leery of its ongoing costs.

From the start of WEAO, the university did what institutions do: assign an advisory committee to monitor the operations and programming. There was much excitement as the first members of the committee met in May 1922, including Wright and Roy Brown, along with the engineering dean and the director of the Cooperative Extension Service. It was the first of many committees established by administrators to provide advice on programming, staffing, and technical issues. The early university committees met intermittently and had little impact on station decision making, but in the years ahead such committees would hold great sway, leading to heated internal battles and public controversy.

This advisory committee's first task was to approve the hiring of the station's first full-time employees: Fay McDowell as radio operator (he earned an annual salary of $1,800) and Ann Charles as program and publicity director and announcer (at $1,500 annually).

THE 'SHOE

Along with the excitement about the birth of a new communications medium in Columbus, 1922 also saw the christening of a massive structure built on farmland near the Olentangy River. Ohio State trustees had voted several years earlier to build a new football stadium, and they turned to the general public for private support. Many pessimists believed the leap from the fifteen-thousand-seat facility at Ohio Field on High Street to a sixty-two-thousand-seat stadium was the stuff of wild dreams. But by 1921 more than $1 million had been raised for the horseshoe-shaped, double-decked concrete structure with bowed and curved sides that would enable fans to be closer to the playing field.

All the excitement surrounding radio had brought Bob Higgy back to campus from his magazine job in Connecticut. His return attracted the attention of the *Columbus Citizen*: "Higgy, though but 20 years old, is considered an authority on radio." With the state funds allocated earlier in the year, Roy Brown and Higgy spent the summer of 1922 planning the first voice broadcasts directly from the new Ohio Stadium. Special telephone circuits were installed from the press box to the transmission facilities through overhead lines. A stadium broadcast booth was padded and draped with studio cloth to make it as soundproof as possible. The station even sent listeners "grid-graph charts" on request so they could track the game while they listened and have a permanent record of the yardage, downs, and scores. The *Columbus Dispatch* reported on the efforts: "Attachés from Station WEAO have set the stage at the stadium and behind the brick walls of the Engineering Lab where the plant is located to attain the maximum efficiency in giving the running story of the conflict and snatches of cheering and band music."

October 7 saw the premiere of the new stadium and of a new form of entertainment for those who couldn't afford the $2.50 ticket. As the Buckeyes beat Ohio Wesleyan 5–0, WEAO broadcast play-by-play at 833 on the AM dial.

The Michigan Wolverines came to town two weeks later on the dedication day for the new stadium. WEAO's technical team had scrambled to expand its signal reach by switching to a 500-watter with RCA tubes. Carl Linxweiler wrote to the station forty years later to describe that day. He was hired as a graduate student in electrical engineering, but before the Michigan game Linxweiler was assigned to be at the microphone for the first half. It was felt that anyone could announce a game if they had a decent voice and an assistant to help with players' names. Linxweiler enthusiastically provided the play-by-play and some color commentary with the sounds of the action in the background. Michigan beat the Buckeyes 19–0 in front of a standing-room-only crowd of seventy-two thousand in Ohio Stadium.

A listener in Stephen, Minnesota, near the Canadian border, sent a note commenting about what he had heard on his radio set that day: "I hope to listen in on more of your programs in the near future and I take the liberty to say that this is the land where the United States No. 1 Early Ohio potatoes are raised." There were also reports of reception from alumni in Chicago and New York.

OHIO STATE DAY

The tradition of Ohio State Day had started in 1911 to reach alumni with entertainment and the latest news from campus. The annual event spawned a series of creative experiments using the latest technology of the times. In 1915 faculty engineers linked the new chimes in Orton Hall to telephones in the Ohio Union to allow alumni in New York and at the San Francisco Exposition to hear Thomas French play "Carmen Ohio." In November 1921, while 8XI was still an experimental radio station, Ohio State Day included a message sent live from the White House by President Harding. It was received by telephone in the electrical engineering lab and broadcast to alumni gathered around radios across Ohio.

The plans for 1922 were more grandiose. With radio operations now fully in place, the Alumni Committee arranged for ninety meetings in locations coast to coast to receive messages and band music via Ohio State's station. J. Lewis Morrill, the ambitious alumni secretary, had secured agreements from stations at the same frequency as WEAO's to stop broadcasting from 7:30 to 8 p.m. for that one night, which would allow a clear AM signal to bounce about the ionosphere and enable alumni listeners with a quality tube receiver and antenna to hear the station hundreds and even thousands of miles away. In a valiant effort possible only during this year of radio development, Morrill received cooperation from stations across the country, helped by alumni in Seattle, Los Angeles, and other major cities who lobbied their local stations to go silent. It was like the station was a commercial clear channel powerhouse for a brief time on a single night. The half hour included a speech from President William Oxley Thompson, music from the University Band and the Girls' Glee Club, and inspirational remarks by football captain Boni Petcoff and coach John Wilce. Ohio State Day of 1922 was a grand success.

The post–World War I boom in radio broadcasting would mark a wondrous period of scientific achievement at Ohio State. The greatest pride belonged to the faculty of the Department of Electrical Engineering, who tinkered and tooled the new medium to life. Building a program schedule and attracting listeners were the challenges of the station's two full-time employees and dozens of students. They would have to be as creative as the engineers, often making it up on the fly as everyone discovered what it was like to entertain and educate using this new thing called radio.

CHAPTER 5

PUTTING ON A SHOW

Regardless of where he may be, the Ohio Stater can tune in at 293.9 meters (833 kHz) and hear the voice of his Alma Mater—the chimes of Orton Tower, a lecture by a favorite professor, football and basketball games.

—*THE OHIO STATE MONTHLY*, 1925

On a typical evening in Columbus in the early 1920s, those scanning the radio dial with their antenna fixed just right and headphones in place could hear many a distant station—but only if those pesky local broadcasters didn't get in the way. In the fall of 1922 radio listeners to the brand-new radio stations in Columbus were lobbying for a "quiet period." They wanted to tune in to the live music, comedy skits, and playlets offered by big-time stations in Chicago and New York City.

In an unusual collaboration, the Columbus radio stations signed a pact to clear the airwaves every evening from 9 to 10 p.m., "so the general public may operate their receiving sets without local interference . . . and to eliminate friction."

Many a listener soon bragged to the local newspaper about holding the record for pulling in the most distant station. Thomas Rife, who wrote a radio column in the *Columbus Evening Dispatch*, often commented on avid listeners: "William Clark, Jr. says that last Sunday night he tuned in KPRC (Houston), KFWB, and KFI in Los Angeles on a three tube set. A half dozen reports from Columbus fans say California stations are coming in without the least bit of trouble."

The silent periods helped audiences across the country enjoy the same radio station, the same entertainers, and the same speeches from political conventions. It was a form of national radio that lasted until 1927, when the commercial networks started connecting stations and listeners could hear New York or Chicago performers on their local commercial station.

With large radio antennas being installed on the roofs of homes across the US to enable listeners to hear stations near and far, it wasn't surprising that there were occasional accidents. A particularly shocking tragedy in Cleveland drew national attention when a fifteen-year-old boy setting up his new receiver attempted to attach the aerial to his home's chimney. He threw the antenna over a high-tension wire, rubbing off insulation from the electric wire. A flash killed the boy instantly, and when his father rushed to try to save him, he also was electrocuted. The Department of Commerce issued warnings about the proper mounting of radio antennas, and businesses cropped up that would install antennas for a fee.

President Herbert Hoover attempted to prod Congress into controlling the unregulated growth of radio broadcasting, but it was unwilling to act. Hoover's Department of Commerce decided to develop a new category of stations based only on technical standards. The idea was that those who could afford to meet high standards would be granted superior frequencies. The new Class B stations required 500 to 1,000 watts of steady power, superior modulation controls, regular maintenance, and quality studio facilities. Most amateur and educational radio stations were left behind as Class B stations became the domain of deep-pocketed commercial enterprises. The new licenses created a superior class of broadcasters who could qualify for special frequencies away from all the clutter.

One intriguing requirement for Class B stations was that license holders could air only live music and information in lieu of playing 78 rpm records. Some historians believe the goal was to counter the proliferation of records by black jazz musicians being played at radio stations. Just a few university stations pursued the Class B professional broadcast standard. One of them was WEAO at Ohio State. By the summer of 1923, director Charles Wright had convinced university leadership to invest in the equipment needed to achieve the highest broadcast standard. WEAO passed an inspection and received the Class B designation, becoming one of only a few educational stations to qualify. The station was allowed to move from the interference-ridden 833 kHz frequency to 1020 kHz, which helped grow listenership immensely. For the next few years, it would be a golden time of broadcasting for the innovative little educational station, drawing notice far and wide.

By the end of 1923 the WEAO schedule included regular market reports, weather forecasts, and news of interest read from local newspapers from 1:30 to 4 p.m. weekdays. The only evening programming was Thursdays with music and educational lectures. The schedule flexed into weekends for coverage of Ohio State sports. WEAO's enhanced broadcast range resulted in piles of letters and postcards from listeners in Vermont, Florida, Texas, and Canada. Some wrote specifically to praise the coverage of Buckeye football and basketball. When WEAO aired several Buckeye basketball games for the first time during the 1923–24 season, the station received a new flood of cards from fans in Chicago, New York City, Boston, Dallas, and Winnipeg.

WEAO began to fill its broadcast time by aggressively recruiting talent from campus and the community. Any form of entertainment was welcome. As an ad in the *Lantern* put it: "If you play the flute, piano, saxophone, harp, or guitar; if you sing or consider yourself fair in giving monologues; if you would like to give a talk on fashion or cooking or decorating; if you are a member of the faculty and want to lecture on branches of science and industry—how would you like to have an audience of from 50,000 to 70,000 listen to you some Thursday evening? You can do all this, thanks to the University broadcasting station, WEAO."

There was no way of knowing how many listeners actually tuned in on Thursday nights, but the promise of such an audience caught the attention of many college deans and student organizations. Program Director Ann Charles tapped into the Greek system and began rotating sororities and fraternities into the studio to present programs. The "TEKE Orchestra" featured piano solos and a chamber orchestra performance from student musicians. Some of the first lectures included "The Prevention of Hog Cholera" from College of Veterinary Medicine dean David White, and "Law, Not War" from Reverend Sidney Sweet, rector of St. Paul's Episcopal Church.

Facing the microphone was a new and often frightening experience for nearly everyone. Some notable OSU faculty sought by WEAO for interviews and lectures refused to have anything to do with the station. The situation became so difficult that President William Oxley Thompson stepped in to issue an edict that all faculty were to "cooperate with the broadcasting station and to prepare programs when requested."

WEAO programming settled into a mix of lectures and live musical performances. In January 1924 the station announced the formation of its own thirty-four-piece radio orchestra of faculty and community members, which performed Thursday-evening programs directed by concert violinist Lillian Belfield.

Several new weeknight programs started airing and would become long-running fixtures. The College of Agriculture developed the *Ohio Farm and Home Hour,* which found a rapt audience among the farm families of Ohio and would be on the station for over two decades.

Local radio audience surveys were completed every year, and the staff paid close attention to the results. With the creation of the radio division of the Bureau of Educational Research at Ohio State in 1931, additional layers of research focused on listenership trends and the effectiveness of teaching by radio.

Because WEAO evolved under the auspices of engineers, the focus at first was on the technology, but with each audience survey, phone call, and postcard, it became obvious that listeners were as interested in the programming as the quality of the signal. In 1924, the tenacity and creativity of director Charles Wright, staff members Ann Charles and Fay McDowell, and student Bob Higgy brought regular live music from an unusual location.

Two sets of wires were run from the station to the iconic chimes in Orton Hall. One line was for the transmission of the music and the other was for communications between the station and the chimes player. For most of the next twenty-six years, Allen McManigal would take a daily break from his job in the engineering drawing department to climb the steps to the tower and play the chimes by manually pushing and pulling levers connected to the twelve bells two floors above. McManigal composed more than three hundred selections using the chimes, and his concerts became a popular radio feature. Taking a lead from the BBC, which used the chimes of Big Ben in London to open its news programming, WEAO signaled the opening of its Thursday night broadcasts with the iconic chimes of Orton Hall.

With the success of connecting the chimes to WEAO, the station started wiring other parts of campus for remote broadcast back to its control room and transmitter. Cables were run through an existing network of underground tunnels to the university chapel, the stadium, and Campbell Hall, overlooking Mirror Lake, where WEAO started broadcasting regular concerts.

A WOMAN'S VOICE

An even more startling sound on radios of the time was the voice of a woman. Radio was a technology dominated by its founders—nearly all men with engineering backgrounds. They were the technical whizzes who found a way to build transmitters and antennas on the cheap, and they were often the first announcers. Ann Charles was WEAO's programmer, publicity chief, and chief announcer, and the latter role gave the station an instantly identifiable sound among the clatter and confusion of frequencies on the dial.

Women had established themselves as Morse code operators before World War I, during a time when the wireless field was desperate for competent telegraphers. Some wireless clubs sprang into existence at women's colleges and as part of the Girl Scouts. A magazine for wireless amateurs profiled Emma Candler, who ran station 8NH from St. Mary's, Ohio, in 1916. The war expanded women's roles in wireless after the navy hired hundreds of young women to train men, who in turn took their skills onto ships. But after the war and with the explosion of radio, few women could be heard on the air.

In addition to Charles, exceptions included Gwen Wagner, who was part of a staff of two at WPO in Memphis in 1922. She was a newspaper reporter during the day and an announcer in the evening. Bertha Brainard was a wartime ambulance driver before she became an announcer at WJZ in Baltimore in 1922. She would rise to become station manager and later work for NBC.

A 1924 radio magazine article took a strong stand against the suitability of the female voice for radio and was backed up by major commercial radio operators. At historic KDKA in Pittsburgh, WWJ in Detroit, and WOR in New York, station leaders said they didn't believe women were fit to be announcers and that a well-modulated male voice was much more pleasing.

FIG. 5.1. Ann Charles with Allen McManigal on piano in the WEAO studio. Charles was also a talented piano and violin player.

The letters and notes Ann Charles received at WEAO ran counter to those opinions, with one writer calling her voice "God given with excellent enunciation." A New York listener found the "lady announcer" very pleasing and proclaimed that "listening to a good announcer is as entertaining as the rest of the program." Jose Velasco, who incredibly received the WEAO signal one night in El Salvador, wrote that he enjoyed the voice of the "lady announcer loud and clear on your marvelous station."

Charles posted a sampling of the many audience letters on the walls of the studio to help with the decor and provide inspiration. There was the card from a blind woman who appreciated how WEAO opened her world to music and information; a note from a six-year-old boy who had built his own radio outfit and loved to listen to WEAO "real loud"; and the letter an Ohio family sent during football season: "Half the town is at the game and the other half is at my house listening in on station WEAO. We'll have chicken dinner ready for your announcer any time he lets us know he's coming over."

That announcer was Lee Henderson, who also taught in the Department of Psychology. The station worked to enhance coverage by linking the announcer booth to both benches, from which field reporters would feed information to Henderson about injuries, substitutions, and other details of the game. It was a sophisticated setup for the times, with eight microphones placed strategically to capture the cheerleaders and the halftime show of The Ohio State Marching Band. *Radio Digest* called the setup possibly "the most elaborate arrangement ever made by any broadcasting station for an athletic event." Photographs of Henderson, Higgy, and McDowell accompanied the article.

The call letters WEAO had been randomly chosen by the Department of Commerce as it hurriedly granted licenses. At the opening of the 1924 football season, the station decided to hold a contest for the best slogan made up of four words starting with W, E, A, and O. Judges included a football coach, a member of the engineering faculty, and a station staff member. The winning slogan, chosen from more than one thousand suggestions, was "Willing, Energetic, Athletic, Ohio." H. M. Young of Uhrichsville, Ohio, received as his reward a pair of tickets to an Ohio State football game. Runner-up choices included "Win 'Em All Ohio" and "Where Education Advances Ohio."

RADIO TELEPATHY?

Innovation seemed an everyday occurrence in early radio. Even educational stations flirted with some outlandish stunts that drew public attention.

Imagine a respected Ohio State professor sitting in a darkened studio with a live microphone and a set of playing cards. Harold Burtt of the Department of Psychology had decided he would test a theory that thoughts could be transferred by radio waves.

Burtt began a series of tests by randomly selecting a card from a deck and telling his audience to think about it. After ten seconds of radio silence in which Burtt himself concentrated on the card number and suit, he asked everyone to write down what card had come to mind. Burtt was also interested in learning how far his listeners were from campus, believing that distance could be a variable in

FIG. 5.2. The first staff photograph for WEAO radio in 1925. Top row: Fay McDowell, operator; Robert Coleman, announcer; Bob Higgy, chief engineer. Bottom row: Allen McManigal, public relations manager; Gwendolyn Almy, piano accompanist; Dr. Charles Wright, director of broadcasting; Ann Charles, program and publicity director; A. M. Ross, agricultural announcer. Higgy, McDowell, and Charles were the only full-time staff.

the transfer of thoughts. He invited everyone listening to his fifteen experiments involving cards, dice, and alphabet blocks to send in their results.

A month later, in a highly anticipated broadcast, Burtt revealed that mental telepathy via radio waves had failed the test and that the number of correct answers correlated with the number of answers one might expect from mere chance or guess. The experiment captured the imagination of listeners and made WEAO the talk of the town.

BY SPECIAL REQUEST

With the ability to broadcast across a great swath of the country, WEAO began receiving personal requests from listeners. The rush of interest started when Mrs. Robert Eakin of Worthington, Ohio, asked the station to broadcast the message that she was trying to find her brother, who had been missing for some forty years. The brother in question apparently heard of his sister's message from the town of Tomato, Arkansas, and immediately telegraphed her. They reunited on New Year's Day in 1926.

It was a well-publicized story of the station's amazing reach and soon WEAO was bombarded with similar pleas. At first, Ann Charles was adamant that all individual requests be aired, but there were no more successful reunions in the weeks to come. Broadcasts about a wife searching for the husband who had fled after two weeks of marriage and a woman who sought contact with the father she had not seen in fifty years proved futile. WEAO soon set a policy that limited such broadcasts.

WRIGHT TO THE RESCUE

Charles Wright was proud of the efforts of the small team he'd assembled to run WEAO, which was complemented by a number of part-timers, volunteers, and students who earned credit through electrical engineering radio courses. In May 1924 Wright sent a lengthy annual report to President Thompson touting the broadcast of 174 lectures (said to be the most of any station in the country), news reports, daily market reports, and musical programs "from the best in Columbus and on campus." He also bragged about the station's play-by-play of Ohio State football and basketball games, the Ohio State Alumni Day broadcast, and special programming during Farmer's Week on campus.

Wright believed that sending reports directly to the president was the best strategy to receive ongoing capital and operating support for the fledgling station. He often advocated for radio broadcasting to become a university-wide function, with funding coming directly from the president's office rather than through the College of Engineering. Among Wright's data was an estimate that one in seventeen Ohio farms, or about seventy-five hundred, had radio receiving sets, and all of them listened regularly to WEAO. With help from the college, the staff had sent interested farmers across Ohio over fifteen thousand copies of a bulletin providing information about buying and using a radio receiver.

Wright also had data from the Universal Radio Time Table organization, which reported receiving more than four thousand requests for WEAO's station listings as a result of an on-air appeal. The organization reported interest from California, Florida, and across Canada, proving the impressive reach of the station's signal.

Wright subsequently requested a new appropriation of $5,000 for specific equipment, including two new 80-foot steel radio towers. His argument concluded that WEAO "extends the influence of a large proportion of the teaching staff to the whole state of Ohio and beyond" and therefore was certainly worth the investment of a prominent land-grant university.

Wright felt he had a persuasive argument. The university had previously invested in WEAO's infrastructure to take it to Class B status, but this time its funding proposal was denied without explanation. In a side note to the dean of engineering, the president took aim at the school's focus on radio communications, calling the effort "something new and novel" and asking the department to focus more on "central things." When the dean shared the note with Wright, he was enraged, ripping off a scathing seven-page response in which he compared the quality of the department's communications coursework to that provided at the Massachusetts Institute of Technology and Harvard.

He wrote that Ohio State graduates were among the most respected radio engineers and communicators in this new field and that he resented that the president didn't appreciate the wide impact of the new medium. It's not known whether Wright's diatribe, with words underlined several times for emphasis, ever reached the president. It didn't matter. The sixty-eight-year-old Thompson soon announced his retirement and was replaced by George Rightmire, dean of the law school. The transition proved to be a positive one for the future of radio at Ohio State.

Wright was constantly concerned that WEAO might lose its important Class B status, which required technical updates. Losing the license would be tantamount to the end of broadcasting at Ohio State, he claimed, and the experience of numerous other educational institutions supported his anxiety. Many university stations had started with great anticipation, only to shut down due to difficulties with funding, federal radio restrictions, or new institutional priorities.

As he prepared his 1926 budget proposal for the new president, Wright outlined yet again the great strides the station had made in distinguishing Ohio State across the state and the country. WEAO, much like the Extension Service, was an important cornerstone of the university's land-grant mission. As radio education scholar I. Keith Tyler later commented about the radio operation: "The people at the station looked upon themselves as an institution which implemented the general policy of a land grant college, which is the borders of the campus are the borders of the state of Ohio."

Wright argued for more operational funding and for the hiring of his resident technical genius, Bob Higgy, who was about to graduate. He wrote glowingly of Higgy's technical prowess and noted that he was also a very good violin player, "making him a combination which is much needed in the engineering of a radio broadcasting station." Higgy had already received multiple offers from stations across the country, Wright added.

To his immense delight, Ohio State's new leadership responded positively. President Rightmire provided the funding to hire Higgy full time and add other personnel and equipment to retain the station's Class B status. WEAO soon moved from two to five full-time staff with the addition of Higgy, announcer Robert Coleman, and a stenographer to help answer all the mail. A new broadcasting station department was created under the dean of the College of Engineering, moving the station out of the domain of the Department of Electrical Engineering. A clear representation of the new importance and respect given WEAO was the construction of two prominent antenna towers at Woodruff and Neil Avenues, connected to a new 750-watt transmitter located in a wooden building built between the towers. Charles Wright's persistence had paid off.

Bob Higgy had been a well-known figure on campus ever since his teen years as an instructor in the World War I–era aviation school. In November 1925 he took to the airwaves himself to present a series of broadcasts about how to get the best reception on radio receivers. It was perhaps the first how-to show on WEAO. As the *Lantern* described it: "[Higgy] is going to broadcast from WEAO an explanation for every trouble, mechanical or spiritual, that you may have in connection with your self-styled super heterodyne radio."

FIG. 5.3. The two towers and wooden transmitter building of WEAO at Woodruff and Neil avenues in 1926.

Higgy's show joined an extensive schedule of educational talks coordinated by Ann Charles: a series of forty-six faculty lectures broadcast from forty different Ohio State departments in 1925. In a *Lantern* profile, announcer Robert Coleman described the programming mix: "Elderly people thank management for old time fiddler's concerts, church music and duets of old favorites. The businessmen on Main Street like to hear lectures, popular readings, and sentimental songs, wives discuss the recipes that are broadcasted. Farmers write to us, vociferous in their praise of the weather reports, telling how they have enabled them to save their hay because they know when to expect rain."

CROSSING A LINE

Charles Wright's advocacy was critical in positioning WEAO for the future within the academic infrastructure at Ohio State, but he also set his sights on the growing Columbus community. In 1926, his aggressive marketing efforts led to a squabble with commercial station WAIU, which complained that WEAO was accepting advertising through partnerships with commercial enterprises. The American Insurance Union had just bought WAIU and would soon broadcast from the Citadel, the downtown forty-seven-story skyscraper the company completed in 1927 (known today as the LeVeque Tower).

WAIU owner John J. Lentz claimed that WEAO's actions weren't consistent with its noncommercial status and that they damaged his station's ability to charge for advertising from the same clients. The complaint mentioned WEAO's broadcasts of performances from groups such as the May Furniture Company Orchestra as an example of indirectly supporting commercial ventures. Wright responded curtly that "the sole objective of WEAO is to give the best possible service to the people of Ohio and for that reason it has broadcast the best talent obtainable," including talent provided through some commercial organizations.

Wright believed WAIU's motive was to have unimpeded access to the best guests and performers in Columbus. WEAO didn't accept funds for services, but it did provide free promotion for local businesses who assisted with programming. Smith Brothers Hardware arranged and supported a program of vocal and instrumental music, and the WEAO announcer credited the company for making the concert possible.

It was a fact that many radio stations were owned by commercial businesses almost solely for the promotion of those businesses, but that was not the model of educational stations. Soon after the WAIU complaint, Wright boldly went a step further and negotiated a deal with the Neil House Hotel in downtown Columbus. WEAO built a makeshift studio in a suite of rooms on the sixth floor of the hotel that was connected by telephone wire to the WEAO campus studio for live broadcasting. The Neil House would receive on-air mentions whenever the studio was used, and the newly created Neil House Orchestra would perform often.

WAIU complained again to the university's leaders, and the issue remained heated for three years until a formal complaint was filed by the American Society of Composers, Artists and Publishers (ASCAP). The union argued that as a nonprofit, WEAO could not claim special exemption from music licensing fees if it was effectively broadcasting commercial programs.

ASCAP wrote that the announcements made after each broadcast were commercial advertising because the announcer would say, for example: "The program during the past half hour was sponsored by the Union Building and Savings Company, located at 48 West Gay Street." WEAO was forced to end its arrangement with the Neil House Hotel and stopped seeking similar commercial partnerships. Wright was bitter about the loss of these community partners, and the station retrenched to the campus. Sponsorship mentions are now common practice in public media, although noncommercial stations have specific criteria they must follow to stay within Federal Communications Commission law.

CHAPTER 6

FIGHTING FOR AIR

> Members of the radio commission assume to tell educational officials of the state of Ohio what is, and what is not, education, and to take from the state the radio facilities used by the state long before the commercial stations were heard of.
>
> —ARMSTRONG PERRY, US BUREAU OF EDUCATION

For the Ohio State radio station, the next few years would be considered an amazing time of audience building with expanded local programming to serve urban and rural listeners. As many as two hundred cards and phone calls were bombarding WEAO every day in 1925 and 1926. Many listeners were curious about the station's program schedule and on-air voices. WEAO decided to produce a report that included a photo of the staff, a basic radio schedule, details about the station's history and operations, and a guide to buying and tuning radio sets for best reception. The station initially printed more than twenty thousand one-page bulletins. Most were mailed through extension services to farmers, and the remainder became a popular item for pickup at the station. The first regular WEAO Radio Program Guide was distributed on November 1, 1926, making it the earliest in the history of educational broadcasting. Other educational stations soon followed Ohio State's model.

Regular programs listed in the first radio guides included *Homemakers' Half Hour* with Faith R. Lanman; *Economic Information for the Farmer*; *Science News of the Week*; *Book Chats*; *Story Hour for Shut-Ins*; *University News*; and *Children's Hour,* with bedtime stories read by Allen McManigal (known on-air as Uncle Al), the chimes player. Daily news features included weather forecasts and

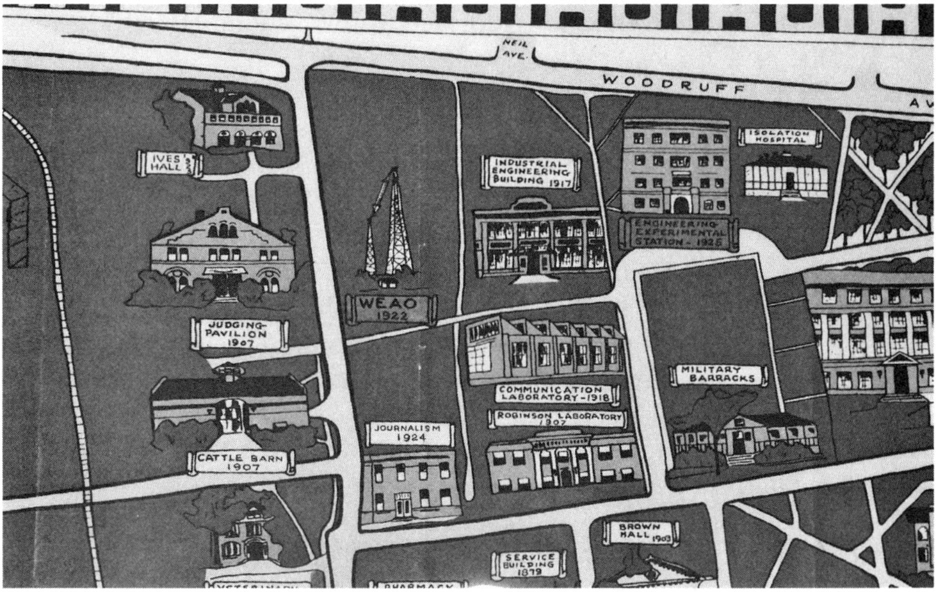

FIG. 6.1. A 1930 map of the Ohio State campus shows the location of the WEAO towers and the Communications Lab studios in the former WWI aviation building.

livestock, grain, and market reports. WEAO partnered with the College of Agriculture to broadcast home study courses to Ohio's farm population. This was the first of many attempts to test the idea that radio could provide the kind of education normally presented in the classroom. Farmers enrolled in the radio courses, received course materials through the mail, and earned a certification after passing a final exam. Such courses were taught by educational radio operations at several universities during the 1920s.

The WEAO Players, a dramatic group that produced many locally written plays for broadcast every Friday night, joined the WEAO Orchestra as a regular fixture on the schedule. Both groups included a mix of community and campus performers, including some of the staff of WEAO. Among the radio actors on WEAO in 1926 was Milton Caniff, who would become a nationally known syndicated cartoonist. Caniff recalled his Dramatic Club's live presentation of the "The Monkey's Paw," which was serious business to those involved. Caniff said he knew "someone had heard the broadcast" because the *Columbus Evening Dispatch* gave it a positive review. "We were certainly not skilled in the technique of radio, but every actor on the show provided his own sound effects," he recalled with pride.

WEAO's broadcast philosophy outlined in the early program guides emphasized that it is a "station *of* the people of Ohio supported and operated *by* the people of Ohio, *for* the thousands of radio listeners in the state and elsewhere." The policy statement went on to make clear that "time on this station is not for sale" and that the "most obscure artist or lecturer may broadcast if the management thinks that his message will be of value to the public." The station mission, at this early point in its existence, reflected WEAO's intent to serve a broad audience with public service programming well beyond the Ohio State campus.

WEAO's technology innovations were noted in several articles in the *Ohio State Engineer.* One piece went behind the scenes of the 1927 Ohio State–Michigan football game to describe how the station handled its broadcast from Ann Arbor. Four microphones were used to cover the action, the announcer, the band, and the Ohio State cheerleaders, The output of the audio mixer was amplified many times and sent to the telephone line leading to WEAO's transmitter in Columbus. Lee Henderson was at the microphone and Bob Higgy provided technical support. The game marked the dedication of Michigan Stadium in front of eighty-five thousand fans, and the inspired Wolverines beat the Buckeyes 21–0. With funding assistance from the *Columbus Citizen* newspaper, WEAO brought the action back home to a large Ohio audience, including the hundreds who tuned in on the latest radio set at the Radio-Electric Shop in Columbus.

Despite the fact that the Ohio State radio operation had evolved into a powerful educational station, widely considered one of the most impressive in the country, trouble was lurking. The federal government was about to strike a major blow to the station's prospects and to the future of noncommercial, educational radio.

FEDERAL RADIO ACT OF 1927

The feds tried to gain control of the growing number of stations and bring some order to the radio frequency chaos through the Radio Act of 1927. It was the first true governmental attempt to regulate radio, and the years that followed challenged even the strongest of the pioneering educational stations.

The act created a powerful five-person Federal Radio Commission (FRC), which had the attitude that popular radio on commercial stations and networks were its prime customers. The days of the Department of Commerce and Herbert Hoover personally deciding what was best for radio were over. Hoover had believed in the strength of educational, informational, and public affairs programming as "a step toward the realization of the true mission of radio."

The FRC thought otherwise and began to unofficially endorse a commercially supported radio system at a time when Great Britain and many other countries had developed a different vision of the new medium. The founder of the British Broadcasting Corporation (BBC), Lord John Reith, believed radio could be effective in serving citizens only if it was operated without advertising. Listeners would pay for the public service programming through a mandatory license on the use of receivers. With its premiere in 1922, the BBC became a fundamental part of life in Britain; today, it is the largest and one of the most influential broadcast organizations in the world.

As it faced the challenge of restructuring the booming radio industry, the radio commission was given the power to grant and deny licenses and to assign frequencies and power levels for each licensee. Sixty days after coming into existence, the FRC ordered all radio licenses void. It would be a fresh start for radio broadcasting, and some seven hundred station owners were rightly concerned about what that might mean. One early indication of the bias against educational radio was the creation of a number of clear channel frequencies. These special radio stations would be able to blast out a signal at 50,000 watts of power, enough to protect the signal for hundreds of miles and force many minor stations into daytime-only operations. Of the twenty-four clear channel frequencies awarded, all went to commercial operators, including WLW in Cincinnati.

At WEAO, Charles Wright had taken a leave of absence due to a serious illness during the summer of 1926. Bob Higgy took over in the interim, and a year later he became the first full-time director of the station at the ripe old age of twenty-six.

Higgy faced the immediate challenge of applying to the new commission with the hope of obtaining a decent frequency. He

FIG. 6.2. In 1927, Robert Higgy became the first full-time director of WEAO radio. For three decades, Higgy ran Ohio State's broadcast stations.

methodically outlined an argument that included WEAO's technical achievements of consistency of broadcast power and staying on frequency, something that wasn't easy with the early transmitters. He wrote that WEAO had been acknowledged "as one of the foremost stations in the country, broadcasting more lectures and other educational matter than any other university station." Higgy described how WEAO was the only Ohio station to provide complete market reports for the state's farmers and to distribute a popular, free monthly program guide, while averaging twenty-five hours of live programming weekly. Higgy's application was cited as among the strongest of any educational station in the country. From a technical, programmatic, and audience-growth perspective, WEAO had immense potential.

The reaction to Higgy's report revealed much about the FRC's attitude toward noncommercial radio. The commission approved WEAO's application on June 1, 1927, but changed the station's frequency from its exclusive 1020 AM slot to 1060 kilocycles, ordering WEAO to share time with another local station. That station was WAIU—the same one that had squawked about WEAO's bouts with commercialism. Within a year, WEAO was again ordered to move its frequency to share time with WKRC in Cincinnati. Such ad hoc orders were becoming commonplace, especially for educational stations. Commercial broadcast networks were coming into existence, and educational stations were in the way.

WEAO had established a strong reputation for quality programming for a broad audience well beyond Ohio, including the many who enjoyed Buckeye sports, but the changes narrowed the station's coverage area and its ability to have regular daytime and evening programming. In addition, it meant that listeners often had no idea when or where to find the station on the radio dial.

Ohio State soon complained on behalf of the station that WKRC hadn't abided by its time-share agreement, going on the air when it desired and causing interference with WEAO's signal. The appeal brought engineering dean Embury Hitchcock and agriculture dean Alfred Vivian to Washington, DC, to appear before the commission. They touted WEAO's strong local programming, but any appeal related to public service seemed to have little impact on the commissioners. At every step when WEAO appealed, the FRC came back with more restrictions. In this case, the commercial station WKRC was assigned nearly all nightly hours "for more popular entertainment programming."

WEAO and Ohio State were far from alone in being targeted for the less desirable frequency assignments and the requirement to share time. At pioneering educational radio station WHA in Madison, leaders shut the station down completely through the summer and part of the fall of 1928. They were so discouraged by the endless interference on their 900-kHz frequency that they ceased broadcasting to fight for a better frequency and rebuild their transmitter, but by November, when WHA returned to the air, the situation wasn't much better. It now had to share a frequency with two other stations, and one of them, WIBO in Chicago, used its powerful signal to overwhelm WHA's programming during evening hours. In 1929, after complaining to the FRC, the commission offered WHA a new frequency of 940 kHz for full-time use during daylight hours only. The station agreed. Today WHA-AM 940 remains a strong daytime public station, but by federal rules it must reduce its power considerably during evening hours to prevent interference with commercial stations.

A FIGHT IN WASHINGTON

Though frustrated, WEAO fought relentlessly to be on the air during the important evening hours, when most people listened to radio. Higgy worked with the faculty broadcast committee and President George Rightmire to challenge the most recent frequency- and time-share order. The resulting order forced WEAO to move up the dial from 550 to 570 kHz and share time with WKBN in Youngstown.

Some two years into the agreement, the relationship went awry. WKBN, which was using the frequency two-thirds of the time, refused to give more time to WEAO, claiming it couldn't release even "a single half-hour and still properly serve the public interest, convenience and necessity in the Youngstown area." In November 1931, WEAO and Ohio State elevated their case and went to Washington to meet with Ralph Walker, the hearing examiner of the Federal Radio Commission. WEAO was fortunate to have support on campus and at the Ohio Statehouse; many educational stations were unable to afford the trip to Washington and therefore couldn't challenge their commercial counterparts.

After testimony, Walker claimed that programming on WEAO was "dictated by the desires and needs of the University, rather than by the listeners in the community," and that WKBN's programs were more of general interest garnering more listeners.

A perturbed Ohio attorney general Gilbert Bettman entered the fray to appeal the decision, starkly declaring that radio faced "prostitution upon the altar of commercial gain." Organized support came from sixteen statewide farm organizations and educational organizations, which sent letters objecting to the order. Bettman focused on the charge that the station operated out of its own self-interest, countering that WEAO conducted regular surveys of its audience to determine programming. He then slammed the commission for its commercial bias: "This entirely erroneous statement of the examiner is apparently predicated upon the theory that the desire and need of the listening public is for still more programs which are purely entertaining. It is submitted that cultural and educational influences are essential to the happiness and welfare of the public, and are also entertaining."

Bettman cited a survey that showed there were more than one hundred thousand radios in the area covered by WEAO, which provided programming "free of all commercial ballyhoo and advertising." The argument and all the support from groups across Ohio failed to sway the commission, which voted 4–1 against Ohio State's appeal.

The legal wrangling was put to rest for the time being. The time-sharing challenge may not have been successful, but it showed that WEAO had some muscle to flex and substantial support across the state. The larger relief may have been that WEAO had survived at all to that point. In 1928 the FRC had requested information and testimony from stations to show why they shouldn't be abolished. Its decisions would rely solely on the stations' quality of equipment and technical capabilities, thus targeting many struggling educational stations. Of those one hundred sixty-four stations summoned to Washington for hearings, half would not survive the commission's scrutiny.

WEAO was entering a new phase. In its first years of existence, simply being on the air seemed astonishing. Receiving a letter from a listener in another part of the country was miraculous. By the end of 1923 the station was widely known for its public service programming and sports broadcasts on a strong frequency without major interference. Four years later the creation of the Federal Radio Commission essentially neutered educational radio and WEAO.

Meanwhile, commercial networks forged ahead offering programs with nationally known performers of drama, comedy, and music that could be heard across the country on network affiliates.

The Great Depression began to settle in across the country. Despite the many challenges to its existence, some educational stations would continue to rise to the occasion and provide exceptional service to its citizens during the troubled years to follow.

CHAPTER 7

OHIO SCHOOL OF THE AIR

> The central and dominant aim of education by radio is to bring the world to the classroom, to make universally available the services of the finest teachers, the inspiration of the greatest leaders and unfolding events which through the radio may come as a vibrant and challenging textbook of the air.
>
> —BENJAMIN DARROW

Benjamin Darrow was an entrepreneur with a salesman's personality, a dapper gentleman with rounded glasses and country wisdom. If he had an idea, he went full throttle toward achieving it and brought many along with him.

Darrow's gift for gab may have been partly genetic. He was related to the famous, fierce litigator Clarence Darrow, who was also born and raised in Ohio. Although Ben traveled extensively and lived in many places, his base was always the farm where he grew up near Mechanicsburg in Champaign County. He attended a one-room schoolhouse before enrolling at Ohio State. Later, Darrow moved to Chicago after successfully selling one of his many ideas: using the new medium of radio as a teaching tool in Illinois classrooms.

The first "school of the air," as the broadcasts came to be called, was developed by commercial station WLS in Chicago after it hired Darrow in 1924. The *Little Schoolhouse of the Air* was funded by the station's corporate owners, Sears, Roebuck and Company. Darrow, the "radio schoolmaster," shepherded the broadcasts on air. He interviewed and hired local instructors for a small stipend to teach classes from the WLS studio. They broadcast programs in music appreciation, geography, science, and farming.

FIG. 7.1. Benjamin Darrow was the driving force behind the creation of the Ohio School of the Air, the first such school in the country.

Darrow noticed during field trips to schools that children paid more attention when someone their age contributed to the broadcasts, so he expanded the role of students in the programs. Within a short time, the radio lessons had a devoted audience of some twenty-three thousand schoolchildren across Chicago and in farm communities in Illinois, Indiana, and Michigan. But the experiment was short-lived. After a year and a half, WLS and its owners pulled the plug in favor of more commercial fare. The experience gave Darrow a taste of the potential of radio for education, which he confidently claimed would eventually "blow the roof off the classroom and expand the walls to the circumference of the globe."

The benefits of radio programming designed for classrooms were obvious to Darrow: It enabled master teachers to share their expertise with many students simultaneously; it exposed students to world figures and current events; it stimulated students' imaginations with drama and music; and parents could listen from home to the broadcasts and continue the discussions with their children. Darrow called radio the "Electric Chautauqua."

After his Chicago experience, Darrow started promoting the idea of a national school of the air. Despite the success of the WLS project and his natural talent for salesmanship, the new concept ran into dead ends at every turn. His pursuit did provide a connection to the Payne Fund in Cleveland, whose administrators were interested in the potential of educational radio. Darrow soon received funding to bring radio education to Ohio classrooms. Additional support came from John Clifton, director of the Ohio Department of Education, who named Darrow the state's first director of radio.

Darrow went to work seeking radio teachers who had the appropriate "personality and voice qualities" and wouldn't be intimidated by teaching in a studio instead of a classroom. He eloquently described the wonder of using radio waves to introduce students to new experiences: "The microphone is blind. It is a strainer which robs us of the effects obtained by the visual, but one that carries the very maximum of aural or impression by sound. Radio will elevate the power of the ear as an avenue of learning."

The Ohio School of the Air (OSA) was launched from Ohio State's WEAO in January 1929, with long-distance telephone lines connecting the studio to WLW in Cincinnati. Both stations would broadcast the programming, but the agreement with WLW was most important. WLW was a commercial clear channel station of 50,000 watts, and for five years starting in 1934, it had a special license to blow out a tremendous signal from an experimental 500,000-watt transmitter so powerful that farmers living near the transmitter site reported hearing the station through their barbed-wire fences.

The first broadcast of the Ohio School of the Air included an organist playing "America the Beautiful," with students singing

FIG. 7.2. Seventh-grade students in Canton, Ohio, taking notes while listening to a music appreciation course in the 1930s delivered through the Ohio School of the Air.

along from their classrooms across Ohio and beyond. A few months later the Ohio legislature appropriated $40,000 to cover two years of operations.

The OSA schedule was impressive, with daily short programs tailored to the needs of children in different grades. There was storytelling for kindergarten through third-grade classrooms, geography and drama for grades six through eight, and chemistry and French for high school students. One popular feature was the weekly broadcast of current events by Harrison Sayre, co-creator of the Columbus-based *My Weekly Reader.*

Soon an estimated one hundred thousand students in Ohio and nearby states were listening regularly to the broadcasts. Following Ohio's lead, similar radio schools were started in Wisconsin, Minnesota, Texas, and Oregon and in cities including Chicago (WBEZ) and Detroit (WDTR). An article in *Harper's Magazine* stated that "in classroom radio education the most conspicuous success to date has been achieved by the Ohio School of the Air."

Thanks to the resourcefulness and genius of Ben Darrow, the OSA would become the nation's showcase for educational broadcasting in schools. The effort also spawned a popular Ohio State radio gathering and some of the earliest radio research ever documented.

RADIO RESEARCH

The first conference of Ohio State's Institute for Education by Radio was organized in 1930 through the university's Bureau of Educational Research under director W. W. Charters. The institute was originally conceived to support Darrow's efforts in researching how best to teach by radio.

Through the bureau's radio division, faculty and graduate students at Ohio State went beyond basic listener surveys to conduct some of the first in-depth studies of programming and listening habits. The bureau worked closely with WEAO and the Ohio School of the Air, using its data to improve the effectiveness of programming. In the early 1930s, Frank Stanton was a graduate student working on models for measuring radio audiences. Stanton developed new gadgets to put on radio sets to show what programs were being listened to and when. After earning his master's degree and PhD from Ohio State, Stanton became a researcher at the CBS Radio Network and, about a decade later, president of CBS.

Stanton was mentored by the young director of the bureau's radio division, Frederick Lumley, who published dozens of articles in national and international journals and magazines. As one example, Lumley's study of the Ohio School of the Air found that many of the radio talks should "dispense with half the content and use the surplus time to amplify and clarify." His work revealed for the first time that the density of information and the pace at which the teachers spoke impacted learning in the classroom.

In one extensive survey of WEAO listeners, Lumley learned that the prime time for tuning in to radio in central Ohio was between 7 and 9 p.m. The station's most popular programming was related to Ohio State sports, but the *Ohio Farm and Home Hour* and the Friday-night drama productions from the WEAO Players also rated high in audience interest.

FIG. 7.3. An Ohio School of the Air dramatic radio production featuring a variety of sound effects. The man on the left is about to slam a door.

Ohio State's groundbreaking research became widely acknowledged. A book about early educational radio said the university was "setting standards for the profession, while its program experimentation is opening new fields and determining techniques for radio as a teaching device."

A shocking blow to the research efforts came in August 1934 when Lumley went missing during a vacation in Glacier National Park. He had set out alone to hike in a remote area of the park and was never seen again. An extensive search failed to turn up even the slightest clue.

Within a few months Ohio State recruited I. Keith Tyler to replace Lumley at the research bureau.

SUCCESS AND STRUGGLES

The Ohio School of the Air was a success almost from its inception, and Ben Darrow became a popular speaker and advocate for radio education well beyond Ohio. Key to Darrow's success in working with teachers was a manual sent monthly to classrooms that described the broadcast courses in detail and included suggestions for teachers' interaction with their students. The manual morphed into Darrow's first book, *Radio: The Assistant Teacher,* published in 1932.

Despite its popularity, the Ohio School of the Air ran into fiscal trouble in 1931, just two years into its existence. With the full effects of the Great Depression hitting Ohio, the state budget was slashed by one third, and the OSA line item was zeroed. Darrow called in every favor he could and generated a public relations effort that flooded the Statehouse with protest letters. Although the OSA had to struggle through a year on a sparse budget, it managed to survive until the legislature restored funding in 1932.

Several Ohio State faculty had been tapped as weekly radio teachers at WEAO as part of the project. Some, like popular professor of English William "Billy" Graves, were less than enthusiastic. Graves said he found it difficult to read his prose and poetry "into a void" from an empty studio, but soon the response from listeners changed his mind: "I was struck by what seemed the very sincere appreciation of what I was trying to do." Graves's programs received laudatory letters from young students, but also elderly listeners and shut-ins across the state.

Within a few years, a new governor brought new challenges for the Ohio School of the Air. Martin Davey, a tree surgeon from Kent, took office in 1935 and zeroed out all OSA funding. Once again Darrow called on his cadre of supporters, and once again they responded. Among the letters was one from the Stark County Council of Parents and Teachers, twenty-four hundred members strong, saying the OSA was doing "tremendous good in the classroom and reaching homes otherwise inaccessible" to quality educational information.

Despite such support, OSA funding was eliminated in the final biennial budget. Governor Davey also sliced more than $1 million from Ohio State's budget, affecting everything from scientific journals to toilet tissue. Students protested and started a massive letter-writing campaign, and most of the cuts were restored by the state legislature.

In 1936 Davey was reelected by a slim margin over John Bricker. Ben Darrow's role in the grand experiment of the Ohio School of the Air ended bitterly the following summer. With no funding, WLW halted its involvement, and the legislature transferred control of the OSA to Ohio State.

Some scholars mark this as the end of the Ohio School of the Air, but it actually continued in some form into the 1960s. The university shepherded it with limited funding through World War II and into the 1950s and '60s under the leadership of Margaret Tyler, the wife of radio researcher Keith Tyler.

After his departure, Darrow traced the history of the OSA in his book *Radio Trailblazing* (1940), which described the challenges and successes of the school. He used his experience to send a warning to all educational broadcast leaders: "There is a danger that broadcasts supported from public funds may become involved in politics, even though the broadcaster plays no politics. This was true of the Ohio School of the Air. It was punished because it refused to become a tool of an administration."

Darrow moved to Buffalo to work at station WBEN and became a popular speaker. In January 1949, twelve years after the demise of his Ohio School of the Air, Darrow was back in Columbus as a special guest on the WOSU version of the school. He was introduced as the original visionary who brought educational radio to the classrooms of America. Darrow, who had been diagnosed with cancer and was in constant pain, made it through the broadcast. He died a year later on his family farm at sixty years of age.

One of radio's inventors, Nikola Tesla, predicted the new medium would serve as a tool for world peace. Others believed it would spawn a cultural and intellectual reawakening. Ben Darrow believed radio would bring the "world to the classroom, to make universally available the services of the finest teachers." The effort to enhance classroom teaching through broadcast radio and, later, television was hampered by more than funding woes. Challenges included acceptance by teachers and administrators; issues with school schedules; lack of interaction with the broadcast teachers; student boredom; and reception and technical issues. Despite all that, there are thousands of youngsters who attended Ohio schools during the worst of the Depression years and found their education enhanced by the work of Ben Darrow and the Ohio School of the Air.

CHAPTER 8

THRIVING IN THE 1930s

> When people here say "The Radio" they don't mean a cabinet, an electrical phenomenon, or a man in a studio[;] they refer to a pervading and somewhat godlike presence which has come into their lives and homes.
>
> —E. B. WHITE, AUTHOR, 1933

Even though a quarter of the population was unemployed during the Depression years, radio continued to grow in popularity. A radio receiver was one of the most prized possessions of any American. Commercial radio networks were now well established and were turning away from classical music, serious lectures, and dramas to lighter fare such as variety shows and popular music. Comedic vaudeville stars had made the transition to radio, and people tuned in to hear George Burns and Gracie Allen, Eddie Cantor, Jack Benny, and others.

In the fall of 1930 author and humorist James Thurber returned to his hometown of Columbus to deliver an entertaining lecture based on his new book written with E. B. White, *Is Sex Necessary?* Thurber was featured in a special WEAO radio broadcast to discuss the book, which examined the great mysteries of life: romance, love, and marriage.

An even greater mystery was how WEAO managed to flourish during the Depression to become an invaluable enterprise. The station had as strong and serious a programming presence through the 1930s as any educational station in the country. Despite being moved by the federal regulators from frequency to frequency across the radio dial, WEAO firmly established itself at a time of great economic turmoil with programming that attracted a devoted and at times large audience.

In 1930, ten years after its maiden experimental broadcasts of weather and market reports to farmers, the Ohio State station aired more than eleven hundred hours of programming, including a symphony series, dramas, lectures, sports broadcasts, programs targeted to rural and farm life, and one hundred hours of educational programming delivered as part of the Ohio School of the Air.

In 1934 the Radio Junior College added even more broadcast hours of formal education with courses for young adults seeking credit in Spanish, poultry husbandry, home economics, business basics, and more. The Radio Junior College was an extension of emergency school activities funded by several agencies of the New Deal. It broadcast fifty-nine different courses in three years on WEAO and enrolled more than ninety-three hundred students. Listeners of any age who took a radio course could come to Ohio State and take an exam to get credit.

Although many of the courses aired from the studio, WEAO wired a few campus classrooms for broadcast. When Harold Burtt taught his psychology classes each morning during the summer, he did so with a regular roomful of students and five hundred others enrolled from home. Lloyd Cook taught a course in social problems with hundreds listening from beyond his classroom. He received a daily stream of mail from taxi drivers, homemakers, and farmers. A group gathered at Put-in-Bay on Lake Erie to listen and then discuss issues facing their community. As part of federal support, radio transcription discs—16-inch platters recorded at 78 rpm—were made of some broadcasts, giving faculty members the rare opportunity to critique their own lectures.

THE STAFF

The creative initiative to produce such a quantity of programming on a shoestring during the Depression years came from just seven full-time staff members at WEAO:

Bob Higgy, director
Ann Charles, program director

Cecil Bidlack, technical supervisor
David Larrimer, chief announcer
Fay McDowell, operations manager
William Knepper, publicity director
Gwendolyn Jenkins, director of drama

The titles don't do justice to the multiple duties the staff tackled. Not only did Bob Higgy, a brilliant engineer, serve as station director, he also hosted a regular summer radio program, and announced many Ohio State football and basketball broadcasts. Chief announcer David Larrimer was studying for his law degree while serving as a writer and actor with the WEAO Players along with working long shifts as announcer. Gwendolyn Jenkins handled the actors and content for all the popular dramatic productions for both WEAO and the Ohio School of the Air. And then there was Ann Charles.

Charles likened her job as program director to that of "running a three-ring circus." Along with producing regular programming, she supervised all the musical performances, including the station's volunteer radio orchestra. She was a regular announcer, and if a guest was late, Charles would grab her violin or sit at the studio piano to fill airtime.

Charles recalled an instance when a studio drama actually caused the shutdown of the station. In the live production, the hero was cussing out the villain, and a police officer was supposed to appear on the scene and blow his whistle. "Disregarding a warning to keep ample distance between his whistle and the microphone, the actor playing the officer stepped up close and blew a terrific blast," as Charles told it. "The result was a complete cessation of broadcasting activities, and the last word to go ringing out over the air was a powerful expletive." Going off the air in such a way brought a deluge of letters from amused listeners.

Charles's husband, Frederick Charles of the College of Horticulture, was a regular lecturer during WEAO's popular farm programs. After he died suddenly in 1930 of complications from appendicitis, Ann poured her energy into raising their two young sons and her career at the station. She remained an on-air presence until her retirement twenty years later.

Bob Higgy loved to stretch the capabilities of radio technology. On a Saturday in June 1930, more than one thousand people

FIG. 8.1. A 1929 photo of the University Players on WEAO radio.

showed up on the Oval to view the unveiling of the 11-foot, 3,000-pound bronze likeness of retired university president William Oxley Thompson. WEAO was there to broadcast the speeches and musical performances. Working with Eugene Weigel of the Department of Music, Higgy also produced a series of band concerts from Mirror Lake Hollow, with wires strung from the studio and microphones tacked to treetops around the lake.

With the New Deal, federal funds supported WEAO efforts to expand local employment by creating the only WPA (Works Progress Association) Radio Orchestra in the country, which put twenty unemployed musicians to work for a time in Columbus. WPA also supported a dramatic troupe that regularly performed, joining the WEAO Players as some of the most popular programming on the station. The actors tackled versions of Shakespearean comedies and tragedies and aired original plays written by faculty and students. For several years, WEAO sponsored a radio play contest between all of Ohio's colleges and universities.

Dramatic presentations demanded a heavy dose of sound effects, and the actors became expert at improvising. Studio A was about fifteen hundred square feet and included props such as a rack with a screen that someone would hit to produce the effect of wind rattling a door, along with a vinegar-soaked hinge that would creak. Actors pressed their thumbs into cornstarch close to a mic to simulate the sound of footsteps in snow. When the script called for chopping wood, they chopped wood. Imitating animal sounds became a competition in itself. The studio had a grand piano and two microphones, one used by the program narrator and one suspended on a long boom for the performers.

Some of those who stood before a microphone for the first time at WEAO went on to stardom. Such was the case of a young black woman who grew up in poverty in the South. Ruby Elzy was discovered by an Ohio State professor who, on a visit to tiny Rust College in Mississippi, overheard her singing and was struck by her "sweet" operatic soprano. With his assistance, she entered Ohio State in 1927 and made her radio debut on WEAO. Elzy later starred on Broadway, in films, and on a national NBC radio series. In 1937 she sang at the White House at the invitation of Eleanor Roosevelt.

Two themes from the first year of radio's existence at Ohio State continued into the 1930s: programming of interest to farmers and rural families and sports broadcasts. G. E. Ferris came to campus with previous radio experience in Kansas and produced the farm-related programming every Monday. His features attracted a strong audience, with more than three thousand rural families requesting to be on the mailing list for broadcast information about the program.

The broadcasting of Ohio State sports, particularly football, continued to be a boon to the reputation of WEAO. In the fall of 1930, Ohio Stadium had a new press box with four booths for visiting stations. Grant Ward, a former Ohio State player and coach, provided play-by-play as WEAO aired all the home games as part of a small sports network of radio stations that included WLW in Cincinnati and WTAM in Cleveland. A new tradition started in December 1930 with the broadcast of the autumn convocation from the campus chapel. The practice continues today as WOSU-TV broadcasts and live-streams on the Internet all Ohio State commencement exercises.

"FIRE THE COACH"

His nickname was "Sad Sam," due to his cheerless demeanor, but his real name was Sam Willaman. His first four seasons as Ohio State football coach offered little to excite the rabid fans, but in 1933 the Buckeyes started strong before losing to Michigan in the third game of the season. After the loss, WEAO announcer Grant Ward, whose full-time gig was as a sportswriter for the *Columbus Journal Dispatch,* wrote an editorial pointedly suggesting that athletics director Lynn St. John fire Willaman. Upset Buckeye players immediately sent a letter to university president George Rightmire demanding that Ward be barred from any further broadcasts. With five games to go, the players didn't want Ward criticizing the team and its head coach, especially on a station run by the university. Rightmire refused to kick Ward off the broadcasts, and the Buckeyes ended the season 7–1. The only loss was to Michigan.

It would be Willaman's best record, but his teams had failed to beat the Wolverines three of the five times they played. Under pressure from alumni and egged on by Ward, Willaman resigned to take the head coaching job at Western Reserve University. Despite mixed reviews as an announcer, Grant Ward continued to broadcast games for WEAO through 1941 and became an even more powerful voice after he was elected to the Ohio legislature.

ONLY THE FITTEST

During the boisterous early years of educational radio, two hundred two licenses were granted to educational institutions that wanted to be part of this cultural phenomenon. By 1937, however, more than 80 percent of those licenses had expired or been revoked by the Federal Radio Commission, or had been transferred to commercial interests. Just thirty-eight educational stations remained. Although radios were in 90 percent of all households in the country and listening averaged more than five hours a day, only a small percentage of that audience had an opportunity to listen to educational radio. Certainly, some of the educational stations that had gone on the air in the 1920s weren't taken seriously on campuses and went out of business without a whimper. Still, it is eye-opening to survey the many institutions just in the state of Ohio that had initially committed to developing radio, but found supporting the stations impossible to sustain.

In the 1920s the University of Cincinnati, the University of Toledo, Antioch College, Marietta College, and the College of Wooster in Ohio all applied for and received official licenses to broadcast. Within a short time each institution gave its license back or let it expire. In most cases, the issue was the expense of running a serious radio operation. The Depression forced state appropriations to universities to a new low. The financial situation, along with regulatory demands for stations to invest in the latest technology, proved a one-two punch that laid out many a station.

Denison University in Granville was licensed to operate a station in 1923 through the Department of Physics. The station, with the call letters WJD, was built for students to gain electronics experience. To meet the federal radio requirements, the university would have had to spend some $10,000. Instead, it decided not to renew its license.

Wittenberg College in Springfield had been among the first institutions experimenting with radio waves, and its January 1922 license was the earliest issued in Ohio. Despite regular programming and university investments, the station folded after the stock market crash of 1929. Wittenberg literally cashed out of the business by selling the rights of its frequency to a commercial station.

Outside Ohio, even Penn State University, a strong land-grant institution, found the situation impossible. The cost of upgrading its transmission facility as required by the Federal Radio Commission forced the station off the air. The university's president fought to save its license with a curt telegram to the commission: "Pennsylvania State College station WPSC is unable to finance equipment change demanded of Commission. We protest this demand upon public educational institution at a time when public funds are seriously curtailed." A detailed appeal was sent but denied, and the station ceased to exist. Penn State would return to radio with an FM station after World War II.

At the University of Arkansas, station KFMQ was forced to share airtime with a commercial station and broadcast only during one-quarter of the daytime hours and none of the nighttime hours. The station leaders believed that "its arms, its legs and its head had been cut off" by the Federal Radio Commission, and the university abandoned educational broadcasting.

At Ohio State, President Rightmire established dormitories in Ohio Stadium during the Depression to provide reasonably priced room and board to students and then cut his own salary 33 percent, but his radio station was never under serious threat of being eliminated.

Ultimately, the future of noncommercial broadcasting that focused on public service rested on just a few stations like WEAO at traditional state universities, most with land-grant legacies. These stations had the resources and political connections to fight off egregious commercial attempts to shut them out. Among the thirty-eight educational stations to survive to 1937 were several pioneers in pre-World War I wireless telegraphy. Some of the original stations are still in operation despite the ongoing limitations of the AM frequency.

Here are the original licensing dates, original call letters, and subsequent call letters (if changed) of some of the first educational broadcasters:

University of Wisconsin	January 13, 1922	WHA
University of Minnesota	January 13, 1922	WLB (KUOM)
University of Illinois	March 28, 1922	WRM (WILL)
Purdue University	April 5, 1922	WBAA
Iowa State University	April 28, 1922	WOI
The Ohio State University	June 3, 1922	WEAO (WOSU)*
University of Iowa	June 26, 1922	WHAA (WSUI)
Michigan State University	August 18, 1922	WKAR
Oregon State University	January 23, 1923	KFDJ (KOAC)

*WOSU's historic AM radio station was sold in 2011 to a religious broadcaster after the station purchased a second FM station.

These and a handful of other stations would sustain educational radio through extremely difficult periods when the future of noncommercial radio hung by a thread. These institutions and the station leaders of the 1920s and '30s survived to lay the foundation for public broadcasting's future in America.

On September 1, 1933, the Federal Radio Commission granted Ohio State a formal change of call letters. After eleven years as WEAO, the station became WOSU to better reflect the commitment of the university to its asset.

ATTEMPT TO ORGANIZE

One of the most challenging aspects of early educational radio was the lack of cohesion among the stations. With the commercial stations developing powerful networks, educators knew if they had any chance of surviving, they had to organize. The Association of College and University Broadcast Stations was one of the first advocacy organizations, and its first executive secretary was Bob Higgy. The association held its first gathering jointly with Ohio State's Institute for Education by Radio in 1930. Higgy believed that one of the only means to fight off commercial radio was to reserve a portion of the frequencies already allocated for noncommercial, educational use.

The university advocacy group was soon joined by the National Committee on Education by Radio (NCER), whose goal was to protect the rights of educational broadcasting. The NCER was funded in part by a five-year grant from the Payne Fund. Its national committee included Higgy and John Clifton, the state director of education.

The university broadcasters and the NCER came together in a failed attempt to advocate for a bill calling for the reservation of AM frequencies for educational use. Still, this first loosely knit group had started to develop true believers in Congress who saw radio as an extraordinary educational resource that was being wasted.

In 1934, Congress enacted the Communications Act, which called for the replacement of the flawed Federal Radio Commission with the Federal Communications Commission (FCC). Although for-profit broadcasters had become the dominant force in American radio, the educational and nonprofit broadcasting communities mounted a new fight for spectrum rights with the somewhat more sympathetic FCC.

A new coalition of educators argued that commercial interests would overwhelm noncommercial radio unless 25 percent of all broadcast channels was set aside for nonprofit public service use. A Senate bill was brought forward with sponsors. Although the effort was again overly ambitious, the educational stations and their allies put increasing pressure behind the proposal. The president of the University of Illinois declared that the Depression had brought on a "sickness" of the national culture and that the recapture of the radio channels was a "necessary remedy." In hearings, John Harney of the Paulist Fathers described the treatment of noncommercial stations as "beggarly and outrageous . . . in a country whose proudest boast is its devotion to the cause of education."

The bill had powerful opposition from the well-funded commercial interests. The National Association of Broadcasters said the initiative would wipe out a quarter of the commercial stations and the investments made in them, but the FCC proceeded to hold hearings on the reserved-channel idea. Testifying in Washington, Ohio attorney general John Bricker touted the educational value of Ohio State's broadcasts to the greater Columbus community. He brought with him endorsements from women's clubs, chambers of commerce, and organizations across the state. A conservative Republican, Bricker spoke frequently and powerfully about the value of educational stations: "By radio, truth must spread over the land. It must not be the agency of propaganda or selfish power."

Meanwhile, the commercial radio networks gave testimony committing their stations to airing educational and cultural programming beyond anything the educational stations could afford to offer. Commercial stations, they said, would "take care" of the educational needs of the populace.

The FCC's report covered some fourteen thousand pages of testimony. The commissioners ultimately found that the radio system was working just fine and that setting aside valuable channels for noncommercial use was not required. They believed programming in the public interest would be brought about by cooperation between commercial stations and nonprofit organizations.

While educational radio had once again been defeated by the powerful commercial interests, the hearings had provided both a forum for publicizing the issues facing educational radio and valuable lessons about how to organize for similar fights over the next decade. The battles of the 1930s to set aside channels for noncommercial AM radio operators foreshadowed the skirmishes ahead with FM and television.

CHAPTER 9

WWII AND FM RADIO

The radio has added five thousand words to our vocabulary, not including those used when the thing won't work.

—*THESAURUS OF HUMOR*, 1940

In 1938 a radio tower to replace the twin towers built on campus in 1926 showed up on the new Ohio State Golf Course, which was still under construction a few miles northwest of campus. Today, The Ohio State University Golf Course has two courses: the Gray course and the Scarlet course. A good drive off the first tee on the 18-hole Gray course can reach a direct descendant of that tower. WOSU presented a special program to christen the new transmission facilities, including interviews with director Bob Higgy and the engineers who put it into service. The tower became an iconic part of the new course after it opened in 1940, with the earliest maps showing it as a landmark. In fact, golf was a regular part of radio programming on WOSU in the 1930s, including an eleven-part series of golf tips with OSU golf coach George Sargent that aired Friday nights.

As the decade drew to a close, another in a series of radio research studies showed that WOSU aired live music nearly 20 percent of the time. The remainder of the schedule was dominated by dramatic productions, book reviews, educational courses, faculty panel sessions, and broadcasts produced through the College of Agriculture.

The majority of commercial stations in central Ohio became affiliated with one of four commercial networks. The networks provided local stations a schedule full of national programming, including a mind-numbing series of radio soap operas. By 1940, most city dwellers owned a radio, but only 60 percent of rural families had one. With some of its most popular programming targeting farmers and homemakers, WOSU found itself successfully reaching those underserved by commercial radio.

WOSU AT 820 AM

J. Lewis Morrill, who once headed the Alumni Association, became the university's first vice president in 1938. Among Morrill's assigned responsibilities was supervision of WOSU, taking direction away from the College of Engineering. Although WOSU may have appreciated having a more direct line to the university president, the change exposed some uncomfortable internal issues.

To better understand the state of radio on campus, Morrill called for a thorough evaluation of the station. A questionnaire was distributed to the staff, and interviews were conducted with faculty affiliated with WOSU. Although it was an overall assessment, the resulting review was surprising in that it centered on what it called the "mediocre leadership" of Bob Higgy and the resulting internal morale issues. It outlined a plan to offer Higgy an opportunity to improve, but it said pointedly that there remained "some question concerning the Director's capacity to grow in leadership ability with the growth of the station, and lead it to greater accomplishments." Higgy had many advocates on campus and within WOSU, but the report was damaging to his reputation. There would be little follow-up. Within a few weeks of the report, George Rightmire retired as president of Ohio State. Morrill soon left to become president of the University of Wyoming and the WOSU review was filed away in Rightmire's papers, likely never to be seen again.

Howard Bevis was inaugurated as Ohio State's new president in October 1940. Over the next sixteen years, as the university grew and matured under his leadership, WOSU flourished, and Bevis would play a vital role in the national fight to save educational broadcasting.

Shortly after Bevis took office, the university approved a deal that would allow WOSU to stabilize its place on the radio dial. The move came after the sharing agreement between WOSU and

WKBN, the commercial station in Youngstown, became particularly acrimonious. WKBN's owner had grown tired of the fight for airtime and suggested the university consider filing for the open 820 AM daytime frequency. WKBN would support the move with the FCC and sweeten the deal by providing $12,000 for a new transmitter to increase WOSU's power fivefold to 5000 watts. WOSU would have its own frequency at 820 AM but would be hampered by the limits of the daytime-only signal, which was mandatory to protect a clear channel station in Fort Worth, Texas.

Despite the daytime restriction, Higgy encouraged the university to accept the proposal. He could now program the station any time of day, and the listening audience would know where and when to find their favorite shows. The increase in transmitter power allowed WOSU to cover much of the state of Ohio during daytime hours. By March 1941, it was a done deal: WOSU had a new radio home at 820 AM and would remain there for the next seventy years. Other pioneer stations, like WKAR at Michigan State, followed the same process in 1941, becoming a daytime-only station to solidify a frequency for the long term.

A HISTORY OF WEAO/WOSU-ASSIGNED AM RADIO FREQUENCIES

1922	833 kHz (All new stations were assigned the frequency.)
1924	1020 kHz (exclusively WEAO)
1927	1060 kHz (frequency shared with WAIU, Columbus)
1928	550 kHz (frequency shared with WKRC, Cincinnati)
1929	570 kHz (frequency shared with WKBN, Youngstown)
1941	820 kHz (exclusively WOSU, daytime-only mandated by FCC)

WWII AND WOSU

The focus of Ohio State and the world changed with the attack on Pearl Harbor on December 7, 1941. The United States was once again in a global war. Addressing the student body in the university chapel, President Bevis said that "everything points to a long, hard war . . . a mechanized war of tanks, planes, and great ships." He was right. The speech was carried by WOSU and rebroadcast by six other Ohio stations. Government agencies imposed wartime broadcast regulations that included suspending local weather forecasts (unless dangerous weather approached) and announcements of casualties. The War Department also sent WOSU a notice that the station was to go silent when requested by air command, though that request never came.

Much of WOSU's programming during the next four years focused on helping its audience understand the international situation and supporting the war effort. Broadcasts included a weekly fifteen-minute series *Ohio at War,* which brought to the microphone Governor John Bricker and experts from across campus. Regular radio discussions and lectures included titles such as "Economics of War," "Keep the Home Fires Burning," and "The Problem of Gas Rationing." Bevis hosted a series on the university's role in the war, and the station carried programming distributed by the Office of War Information, including the series *Uncle Sam,* with top radio stars of the time providing information on how every citizen could "help win the war."

For comic relief, the campus once again welcomed back writer James Thurber at a student rally in March 1942. Thurber, who had attended Ohio State for several years but never graduated, was interviewed by WOSU and then joined the crowd at the RKO Palace Theatre to enjoy the debut of the movie based on his play *The Male Animal,* starring Henry Fonda.

The WOSU Players presented an original patriotic play that year by student Leonard Wolf that told of the spirited struggles of the people of Europe against their Nazi occupiers. WOSU's programming director, Ann Charles, organized a drive for 78 rpm recordings of light classical music and musical comedy to be collected at the station and sent to military hospitals ("no swing or heavy classical records desired").

And music continued to be a fixture on WOSU, with one special program on the life of Italian tenor Enrico Caruso that included announcer Ed Sprague singing arias. Sprague had been brought in by Higgy to fill some of the wartime vacancies. The staff nicknamed him "Pappy" because he had been at Pittsburgh's KDKA during the birth of radio broadcasting decades earlier.

Among the staff members who left to serve in the war were announcer Casimer Bielski (who used the on-air name Fred Bielson) and production supervisor Friel Heimlich, both of whom were called up from the reserves to active duty and shipped overseas. By war's end, Colonel Heimlich, fluent in German, was serving as an intelligence officer for the US forces in Berlin. Heimlich became an avid collector of war memorabilia. Among the souvenirs he sent to his wife in Columbus were Hitler's opera hat, which he claimed had been found in the Reich chancellery air-raid shelter in Berlin; a Nazi party badge belonging to Hitler associate Joseph Goebbels; a huge swastika banner that had hung in Berlin; and one of the medals Hitler awarded to mothers who bore six or more children to serve in the Reich. For a time, Heimlich's loot was part of an exhibit in the Ohio State Archaeological Museum.

Margaret Tyler, director of the Ohio School of the Air, recalled an early-morning call on June 6th 1944 during the war: "Our telephone rang at 2 a.m. and the whole staff was called over to WOSU because the invasion of Europe was on. We worked all night (reporting the news) and all the next day without stopping. They sent food in to us. I tried to monitor Radio France with my very poor French to get the word as it came direct from Paris. It was an exciting time."

Wib Pettegrew, a former drum major for Ohio State, served in various positions at the station starting in 1934: radio playwright, actor, and director; sports play-by-play announcer; and production supervisor of Radio Junior College programs. During wartime, he was promoted to supervise all WOSU radio programming. In 1943, he narrated a motion picture called *The University and the War,* which depicted Ohio State's role in the war effort.

Pettegrew wrote, produced, and announced a weekly program for grade school students that attempted to explain the situation in a conversational manner, as in this excerpt from one of his 1945 scripts: "Hi there, boys and girls. If you have a calendar hanging in your room, take a look at it. Under tomorrow, Friday, you'll see a big number SEVEN. December 7th is the anniversary of Pearl Harbor, the day the Japs attacked us and drew us into a war. That was back in 1941—four years ago. The war is now over. The Japs started the fight, and we finished it. This year, on the anniversary of Pearl Harbor, we're not thinking much about the war. No, on this fourth anniversary of Pearl Harbor, we are thinking mostly about peace—and how to gain peace."

As the war came to an end, Margaret Tyler hired graduate student David Ayers to direct weekly radio dramas for children as part

FIG. 9.1. Wib Pettegrew worked with returning veterans to produce a series of radio programs about their experiences overseas and back home in 1943.

of the Ohio School of the Air. They included *Boys and Girls in Bookland* and a historical drama series, *Once Upon a Time in Ohio.* These fifteen-minute programs used young student talent from the College of Education's experimental K–12 University School at Woodruff Avenue and High Street.

Robert Butche was one of Ayers's student actors from the University School. Butche would take the streetcar up High Street from his home in the neighborhood now known as the Short North to attend school; then he would walk the two blocks to the radio station on 19th Avenue. Butche said youngsters had to be punctual and ready to perform at WOSU, and there was little time for fun in Bob Higgy's studio. Butche was involved in some forty broadcasts from Studio A and another dozen from the smaller Studio B during his years at the University School. As Butche grew older, his respect for Higgy and the professionalism of the station grew: "Higgy was a rigid guy. Intent and proud of the institution, and he took it damn seriously and he didn't want people around who didn't take it seriously. Here is someone who did something really remarkable and for many years. There was no horseplay. This was a very straitlaced professional place."

FIG. 9.2. Students from the experimental K–12 University School would walk down 19th Avenue to WOSU to perform in dramas on *Ohio School of the Air* and their own radio show called *Buckshot*.

University School students wrote, acted in, and produced *Buckshot,* a regular variety show on WOSU that included sports, news, and drama segments. In the early 1950s, a teenaged Bill McCormick was the primary announcer for some of the shows. He loved that the station experimented with dramatic productions. “They would put me on the air as a stagecoach driver one week and a bank president the next,” said McCormick. “Students were writing scripts right and left.” As a teenager, McCormick worked at WBNS, the

CBS affiliate in Columbus, with comedian Jonathan Winters, who started his broadcast career there.

FM HITS THE AIR

FM radio, as demonstrated by American inventor Edwin Armstrong, provided high-fidelity sound far superior to AM radio's sound. The Institute of Radio Engineers, the organization created by Bob Marriott, met in Boston in 1938 to hear an Armstrong demonstration of FM. It was the largest gathering in the history of the institute. Newspaper reporter Henry Lane of the *Boston Sunday Post* described the scene: "Virtually spellbound, nearly 600 college professors, engineers, technicians, scientists and the curious sat for well over an hour listening to all manner of program material. Static, tube hiss, hum and the other distracting sounds that nearly always accompany conventional radio reception is entirely lacking." The system, Lane wrote, "will reproduce silence itself."

Armstrong lobbied the FCC for a small portion of frequency to test FM radio. The FCC granted his request, and he built a network of FM stations in New England. Compared to the sound quality of static-filled AM radio, the FM high-quality sound proved the worth of the new FM band. The World War II years led to the further refinement of FM technology. Higgy wrote in 1944 of the difference in quality between AM and the coming FM frequency, making an interesting point that frequency modulation was "a new type of short wave radio broadcasting that provided excellent service over limited local areas." This allowed a great many potential FM stations in a market because, unlike AM broadcasting, there would be less interference on similar frequencies.

In September 1944, with the war still raging in Europe, the FCC began a series of hearings on the allocation of frequencies for FM radio. It was a crucial time for the future of educational broadcasting. Thirty educators testified in Washington, urging the commission to reserve frequencies. Among those testifying was OSU President Howard Bevis, who represented the Radio Broadcasting Committee of the National Association of State Universities. Bevis was encouraged to speak after receiving a note indicating that FCC member Clifford Durr wanted the institutional leaders from Ohio State and Wisconsin, the two strongest educational radio stations in the country, to appear before the commission. Durr, of Alabama, was a champion of noncommercial radio, and, later, civil rights, who saw the coming of FM as an important opportunity to develop an alternative system to commercial network radio.

During his testimony Bevis said FM channels were needed to augment the struggling AM stations and extend local public service coverage beyond daytime, which severely limited programming and listenership on AM stations like WOSU.

Unlike their rabid opposition to reserving AM frequencies for educators, commercial broadcasters offered muted resistance to the call for FM channels. Given the lack of FM radios in home or cars, they simply didn't see FM as a major threat to their profitable operations. For about three decades they would be correct. In June 1945 the FCC announced that twenty FM channels from 88.1 to 91.9 MHz on the FM dial would be reserved for educational purposes. Educational broadcasters rejoiced. For those who were tired of the commercial domination of AM radio, this was "radio's second chance."

AN OFFER FROM STANTON

After he testified before the FCC in favor of the allocation of frequencies for educational radio, Bevis received a lengthy letter from Frank Stanton, the vice president of CBS Radio, who had both a master's degree and a PhD in psychology from Ohio State.

Stanton's letter inquired about the role the administration thought radio and communications could play in the postwar future of Ohio State. It outlined the many disparate areas of expertise in the field that gave the university the potential to be the "foremost educational center in the world for broadcasting."

Stanton accurately described Ohio State's history of contributions to the broadcast industry. He said the annual Institute for Education by Radio was the "only forum for serious discussions about radio's public service." He lauded the annual radio engineering conference on campus, the operations of historic WOSU, and the distinctive radio research unit he had been involved with as a graduate student. His suggestion was to integrate multiple disciplines across departmental lines and merge journalism and radio plus research in motion picture techniques into a powerful new communications school. Stanton believed Ohio State's unique place in the history of radio would fade if the university didn't capitalize on it. The time to act was now. He concluded with a commitment to "help in any way to implement" a plan if there was interest. A year later, Stanton was president of CBS. He would become a central figure in the development of television in the United States. No response to Stanton's offer could be found in Ohio State's archive.

WOSU TURNS TWENTY-FIVE

As the war wound down in 1945, Ohio State entered a time of unprecedented growth, with double the expected enrollment during the next year. WOSU celebrated its twenty-fifth anniversary as a radio station with a special broadcast including messages from Howard Bevis and a symphony concert. Bevis emphasized the broad mission of the educational station as part of the founding of Ohio State: "From their very beginnings, back in the days of the Civil War, the great land-grant universities such as Ohio State have had an obligation to serve every citizen of the state." Bevis proudly pointed to the thousands of students across Ohio being educated by WOSU programming without ever seeing the campus; the broadcasts to farmers with information to help them produce the vast amount of food needed during wartime; the promotion of war bonds; and WOSU's public service efforts to encourage citizens to contribute to the Red Cross.

Serving Ohio's farmers continued to be a key goal of WOSU programming. A collaboration with the Ohio State Extension Service had led to the purchase of the station's first remote truck in 1944. It was used by host and agriculture expert Bill Zipf to travel into farmers' fields and make recordings on a magnetic wire recorder, one of the earliest versions of a reel-to-reel tape recorder. The truck was outfitted with a shortwave transmitter that had a range of about 15 miles, allowing for periodic live, in-the-field broadcasts direct to WOSU's studios. The recorder weighed some seventy pounds and required an audio engineer to run it, according to Al Biogonni, who

FIG. 9.3. WOSU farm editor Bill Zipf interviewing a farmer atop the station's mobile production truck, acquired in 1944. The truck had a platform that was used primarily to announce OSU sporting events.

would be a radio engineer at WOSU from the early 1940s into the 1970s.

The recording of local radio programs for replay on stations remained a major issue. Most stations, including WOSU, had acquired disc recorders, but creating a master record of a local program was laborious and the results were often distorted. One of the problems of disc recordings was that when the playback needle hit a flaw in the groove, it repeated the same sounds endlessly. At KWSC, an educational station at Washington State, a portion of a disc recording of a US senator had the politician saying: "My record is clear on Grand Coulee Dam. The voters know this." What came to air due to the flaw in the groove was "dam the voters . . . dam the voters . . . dam the voters" until a student operator ran back to the control room and pushed the needle forward. The magnetic tape recorder, developed in Germany, revolutionized radio broadcasting, but it wouldn't become commonplace at stations until after WWII.

In that 25th anniversary speech, Bevis brought some seemingly great news to WOSU, predicting developments ahead in radio and television and "the potential of new studios for WOSU to keep up with the times." It seemed the station might soon be able to exit its old WWI facility for new studios. A few months later the university's long-range building plans were released, and for whatever reason, WOSU was left out of the plans. Higgy promptly begged Bevis for upgrades to the radio complex. He described the space as "horribly inadequate" and asked for funds to provide professional sound-absorbent treatments for the studios, more shop space to fix equipment ("the current space is often near freezing during winter months"), and new heating systems ("the radiator bangs loudly during broadcasts"). His appeals went unanswered, and soon Higgy's tenure at WOSU was again under fire.

KEITH TYLER

A thin-faced, balding Nebraskan named Keith Tyler had come to Ohio State in 1935 as a professor and research associate in the Bureau of Educational Research. Tyler was a minister's son, a smart, thoughtful, highly skilled communicator. His job was to bring new research to the field of radio education and expand the regular gatherings called the Institute for Education by Radio, which W. W. Charters, his boss at the bureau, had started. With Tyler's influence and determination, the meetings became a prominent think tank for both commercial and noncommercial radio. The institute published a book every year that included the speeches and papers presented during the proceedings.

Many believed the institute was one of the leading forces in the history of American broadcasting, with regular presentations from the likes of Edward R. Murrow and members of the FCC. The week-long meetings gave those in the educational broadcast field an opportunity to share and to build relationships. "Tyler was the guy who carried it on with brilliance and sweat and style," said pioneer television broadcaster Dick Hull. "It was really an Academy. The meeting place for all of broadcasting in the United States.

An early radio history recalled the pub in the Deshler-Wallick Hotel downtown, where the real business of the institute took place,

FIG. 9.4. Keith Tyler would be a dominant force in educational radio research and development by hosting major institutes and developing the first broadcast awards program. His wife, Margaret, would run the Ohio School of the Air in the 1940s and '50s.

and the Gayety Theatre on High Street, where stripper Rose LaRose always seemed to be performing at the same time the institute had gathered hundreds of broadcasters.

Tyler had the foresight in 1936 to create the Ohio State Awards, the earliest national awards in broadcasting for serious radio programming. The coveted honors went to children's programming, in-school programs, news and public affairs, agricultural programs, and more. The concept was to have an exhibition of programs at each institute from both educational and commercial broadcasters and to call attention to merit without making it become a bloody competition. Indeed, through the decades the Ohio State Awards continued to be highly treasured even after the Peabody and Emmy awards were established.

After World War II Tyler was hired by the War Department to help educators in Germany rid their classrooms of Nazi influence. Hitler had used school radio broadcasts and films to get his propaganda "into young people with youth movements, organizations and Nazi groups," Tyler said. He recalled sitting on a panel in Berlin with representatives from other countries to review the films used in German schools. "Our job was to censure out any evidence of the Nazi movement," he said. "If we saw young people in Nazi uniforms, that was cut out or the whole film had to go."

By 1947 Tyler was back at Ohio State, where his research into the effects of radio programming on children and the role of radio in education led to his being assigned to direct national studies funded by the Rockefeller Foundation. Like Ben Darrow years before, Tyler would become a forceful voice for the use of the medium in formal education.

President Bevis named Tyler to the new campus position of director of radio education, and Tyler used that post and his scholarly influence to advocate for the creation of a powerful group of faculty to "properly integrate radio activities on campus." The Radio Advisory Council's original charge was to approve all instructional courses broadcast on WOSU, review all equipment requests related to radio instruction, and facilitate a smooth working relationship between the station and academic departments.

Tyler soon expanded the council's oversight to overall operations, including the working conditions and morale of the staff. As he put it in an oral history: "WOSU was flourishing and Robert Higgy was the director. He had come up through engineering and was a conscientious, hard-working person, but his background lacked something in the field of programming and education. Here's an educational station run by an engineer. In my way of thinking, that didn't reflect the kind of leadership that ought to be offered."

Higgy met with Harvey Davis, vice president in charge of faculty and instruction, to express his objections to the council's charge. While low morale wasn't unknown at WOSU, Higgy countered, the work environment and an overworked, underpaid staff naturally led to such issues.

The members of the Radio Advisory Council would be a force in the evolution of WOSU radio and television into the 1950s. The council included the following people:

- Keith Tyler, chairman and internationally respected researcher in teaching by radio (his wife, Margaret Tyler, was director of the Ohio School of the Air)
- Harrison Summers, professor of speech communications who had worked for several national radio networks in audience research
- Ken Dameron, professor with a wealth of publications in the field of advertising
- William Ewing, program director for WOSU radio and instructor in speech communications
- Eugene Weigel, longtime director of the School of Music
- Paul Wagner, professor in the School of Journalism
- Robert Higgy, director of WOSU radio since 1927

Although he was a WOSU staff member, Ewing aligned with Tyler's goals and wrote numerous notes to him about the radio operations. On the subject of a new project, Ewing wrote that Higgy had "the same old penny pinching reaction. No doubt he will deliberately sabotage the whole thing." He stated in one note that he was tired with dealing with Higgy, but that he "was in a fighting mode."

A 1948 press profile of the station was all positive until it came to a quote from Higgy, who was once again sending a message to the university administration about the lack of support: "We have very poor studios, probably the worst in the state. Outside noises get into the broadcasts and there isn't anything we can do about it under the present conditions." Bevis was able to ignore much of the drama at WOSU because he delegated power to the Radio Advisory Council to coordinate Ohio State's entire radio program. The most important decisions related to running WOSU were now in the hands of Keith Tyler.

FIG. 9.5. The August 1949 WOSU program guide. WOSU radio was the first educational station to provide an audience guide in 1926.

WOSU-FM ON THE AIR

A few months after the allocation of frequencies in 1945 for noncommercial FM radio, Higgy filed a report with Bevis outlining plans for development of an Ohio educational FM network of five stations. The idea was that Ohio State would serve as the hub and the other stations would be located at four Ohio universities in Bowling Green, Kent, Oxford, and Athens. The stations would be networked together to provide an efficient model of broadcasting. Higgy believed that the network could reach some seven million residents. Bevis did not respond to the proposal. The year before, an even more ambitious attempt by the State Department of Education had called for a network of sixteen FM stations in Ohio to transmit educational programs to K-12 schools and colleges, but the million-dollar price tag was too much for the state.

There was similar activity in Wisconsin, where a state radio council was created to investigate the creation of an FM radio network. In 1947, WHA-FM went on the air in Madison, and the next year state funds were appropriated to start building more stations. The eight-station network was in place by the early 1950s. Due to continuous funding challenges and political issues, no statewide FM network was ever developed in Ohio.

Simply getting one FM station on the air at Ohio State was found to be challenging enough. Although Bevis helped lead the charge for the set-aside frequencies, Ohio State lagged in applying for an FM license. After years of back-and-forth about cost estimates, the trustees eventually approved the license application, and WOSU-FM 89.7 finally went on the air December 13, 1949, with 50,000 watts of effective power.

The FM transmitter and antenna were positioned next to the AM tower on the university golf course, and the station started simulcasting the programming being aired on the AM station. WOSU-FM was one of ninety educational stations to come into existence on the new band in 1949 and 1950. It joined three other Columbus FM stations already on the air: WCOL, WELD, and WCLT. The stations didn't have much of an audience, since radios were rare and priced at many times the cost of AM radios.

The first move by Tyler's Radio Advisory Council was to start broadcasting the AM and FM stations on Sundays, when studies showed substantial radio listening. This was a first for WOSU, which rarely broadcast on weekends at all and still turned off its transmitters during holidays and other times when the university wasn't in session. Tyler also decided to focus on developing radio journalism at WOSU. The station was known for hiring announcers with a background in theatre who regularly read national and state wire copy. Topical programs were generally presentations from professors giving perspective to issues of the day.

An incident accelerated the need for internal journalistic expertise at WOSU. The station was roundly criticized by listeners for airing a program from Ohio State's Department of Economics in which a faculty member tried to use a local analogy to illustrate the difference between communism and socialism. Columbus's water department might be considered socialism, he said, because it was financed by consumers. However, the police department was more like communism, because it was financed by the government. The chair of the economics department responded to listener complaints by saying the program wasn't "good radio," but noted that "criticism like this is to be expected . . . as some listeners will be lacking in scholarly competence."

WOSU leadership realized they could no longer schedule programming from just any academic area without much thought. Station leaders had to take ownership of everything they aired, no matter the source. Higgy responded to the criticism by deciding to cease all broadcasting of political news, and the order was posted at the station. The Radio Advisory Council disagreed with that approach and lobbied for Paul Wagner to work as part-time news director while also carrying his teaching load in the School of Journalism. The council would always prevail over Higgy. Wagner was allowed to hire an assistant to supervise the news more closely. He also drafted the station's first news policy—which was approved by the council without Higgy in attendance—and began to build a stronger student presence in radio news.

WOSU started tackling serious issues through national programming received from large metropolitan stations such as WNYC in New York City. The new journalistic content included a fourteen-week series on venereal disease and one on issues facing minorities. As the station matured in its coverage, program director Bill Ewing began to feel some heat. Academic units were used to getting regular airtime to showcase faculty expertise, but Ewing had established policies limiting each department's access. WOSU's reach to 1.2 million citizens across Ohio meant that "the station must present radio that is of broad interest to that population," he said, "instead of serving as a publicity arm" for university departments.

Wagner was auditioning and hiring students to announce the news, but his efforts were met with some disdain. Listeners criticized the students for their pronunciation, their selection of news items, and their "slanted" reports. The complaints found their way to the president's office, resulting in an informal ban on student news announcers, though they could work behind the scenes at the station.

One of Ohio's most prominent politicians and a member of the university's board of trustees, US senator John Bricker, had received a great deal of positive exposure on WOSU radio during his political career. Now, in the 1950s, Bricker was obsessed with the potential influence of communism. He had asked the FBI to investigate certain faculty members he thought were communists. (The agency refused.) Bricker started to send notes to Bevis objecting to certain WOSU programming, saying: "I don't see why Ohio State has to support radio programs such as indicated on the enclosed sheet and especially at a time like this when our country is endangered."

The list included programs distributed through the National Association of Educational Broadcasters, including commentary about "the real issues of our time" and a program about "the race problem" from the secretary of the NAACP. Bevis explained to Bricker that some of the "objectionable material" had already aired, although the program about racial issues was shelved as propaganda. "We try to watch these matters, but this time we slipped," wrote Bevis.

Bricker soon was questioning other programming content and taking serious interest in every listener complaint that came to his attention. In one case in April 1952, Bricker requested transcripts of a WOSU program called *Background of the News,* prompting the *Lantern* to write: "Now we've heard everything! WOSU has been charged with broadcasting 'biased, anti-Republican, and left-wing' news. . . . That's almost funny. . . . Times being what they are, one complaint means another inquiry. Today, it just doesn't pay to be controversial." Within a week, Bricker cleared WOSU of the complaint, saying the program was a "fair representation of the news."

Such was the tenor of the times that a US senator would spend his valuable time reading a dozen transcripts of radio news shows in response to a single complaint from a listener claiming anti-Republican bias.

The Radio Advisory Council continued to overrule Higgy on programming decisions, choosing to air some programming that the council knew might be controversial, such as "Lest We Forget," a national radio program about race relations. The council sought the backing of university administration to present a united front in support of the broadcast, but Ohio State declined to take any stand on the issue. Council chair Keith Tyler believed such programming decisions began to strain the council's relationship with university leaders.

THE SUMMERS STUDY

Harrison Summers, Radio Advisory Council member and speech communications professor, was working with students to develop and administer multiple audience studies focused on WOSU. In one case, they conducted an extensive house-to-house and telephone survey in twenty-two counties within a 40-mile radius of Columbus.

Summers and his students learned that WOSU had "prestige" in the community, but actual listening was much less than expected. They also found that although WOSU's focus was on serving people with college degrees, roughly half of the adults in the listening area had less than a high school education.

To build audiences among both groups, Summers advised the station to offer a substantial number of programs meant to entertain as well as educate listeners. In addition, he recommended that WOSU consider block programming in order to expand listenership. One block of airtime might target homemakers, with another block devoted to farm listeners. Newscasts would act as a transition between blocks. It was a suggestion that WOSU and stations across the country would begin to adopt.

Summers also believed WOSU could gain a loyal audience by developing "personalities" identified with the station, pointing to Chet Long and Irwin Johnson of WBNS and Ruth Lyons of Cincinnati's WLW as examples. Finally, he wrote, WOSU should "never be

FIG. 9.6. Marv Homan (right) providing color commentary for a 1952 OSU football game on WOSU alongside Bob Holsinger. Homan was the voice of the Buckeyes for over three decades.

characterized as an 'amateur' or student station, one whose standards were lower than those of competing commercial stations." In one reaction to the report, the staff voted unanimously that the station discontinue using a radio announcement calling WOSU "your station for education," which stereotyped the broadcaster.

The station wrestled internally with the potential of broadcasting more popular music than the typical classical fare. Some thought providing "more accessible" music programs might bring a greater audience to the rest of the educational programming schedule.

FIG. 9.7. WOSU radio program director Don Quayle conferring with a University Hospital physician about a future program. Quayle would become the first president of National Public Radio.

WOSU IN THE 1950s

With the growing emphasis on news coverage, Robert Holsinger became WOSU's first full-time news director in 1951, while also assisting with the play-by-play of OSU sports on the station. A special series that year focused on the need to educate audiences about American values and the democratic way of living. It was safe radio. In one broadcast, Bevis pointed out that studies showed an "appalling ignorance of the basic concepts and facts about America," including little understanding of the First Amendment to the Constitution.

The journalistic efforts were given greater emphasis with the arrival of program manager Don Quayle and newsman Al Hulsen in 1956. The two created a true professional newsroom, encouraging staffers to pursue local stories in the field and to bring newsmakers to the studio for interviews. This was new territory for WOSU, whose announcers traditionally read United Press International wire copy or excerpts from Columbus newspapers. WOSU was now actually covering and breaking the news. In its first experiment with a news call-in program in 1959, the station was overwhelmed by the response. As newsman Don Davis put it: "We usually didn't know if anyone was listening."

Don Quayle grew up in Utah and as a youngster listened to the Ohio School of the Air. "The possibility of working at WOSU was big time," he recalled in an oral history. In January 1959 Hulsen and Quayle reformatted the 5 p.m. drive time by taking four fifteen-minute programs and turning them into a seamless one-hour newsmagazine called *On the Way Home*. Announcer Gene Gerrard hosted the instantly popular show, which combined local interviews with news, sports, music, and information such as travel and cooking tips. Homebound motorists could hear the news along with lighter segments such as "Leisure Time," hosted by Bob Schweikart and Olive Haynes, which focused on events and activities in the Columbus area.

When Quayle left WOSU in 1960, he joined WGBH in Boston for a few years. He then ran the new Eastern Educational Radio Network, where he again teamed with Hulsen to create a daily news program called *Kaleidoscope*. In 1970, Quayle became the first president of the new educational network called National Public Radio. He hired Susan Stamberg, who helped create an afternoon drive-time series in many ways similar to the program Quayle had tested at WOSU.

All Things Considered premiered on May 3, 1971, as NPR's first national news program and was originally picked up by ninety educational radio stations, including WOSU. It was hosted by Robert Conley; produced by Jack Mitchell, who would later run Wisconsin Public Radio; and delivered to stations by a telephone line. Quayle wanted *All Things Considered* to be the public radio version of *Sesame Street*, that is, a program that would define NPR and bring recognition and attention. In 1979, as more educational stations started broadcasting in the mornings, NPR introduced another national drive-time program, *Morning Edition.*

IN THE BOOKSTALL

"Welcome to the *Bookstall.* Today I'm going to continue reading to you from the memoirs of Sherlock Holmes from Arthur Conon Doyle." So began Gene Gerrard announcing another *In the Bookstall* program in 1977. Twenty-seven years earlier, the WOSU program started airing for a half hour daily at 1:00 p.m. with Wilson Dumble, a retired assistant professor of English, reading well-known plays, books, and poems. Dumble had worked in Paris during World War I as a journalist for *Stars and Stripes,* the armed forces newspaper. His program became a huge hit, especially among homemakers and the elderly.

For a short time, Dumble was one of the station's most popular personalities, although he occasionally sparked controversy. He once read a new play called *The Moon Is Blue* that included the word *hell* sixteen

FIG. 9.8. Gene Gerrard would elevate the popularity of the daily *In the Bookstall* series, which aired on WOSU for over three decades.

times. The audience and Dumble's bosses were not amused. When he retired in 1958, his successor, Gene Gerrard, developed an even more loyal listenership. Gerrard was already well-known in Columbus as a theatre critic for a local newspaper and announcer on WOSU radio. He performed in plays at the Stadium Theatre on campus and had a show on the new WOSU-TV station called *Pronunciation Please,* which included a segment in which contestants would hear a voice and try to identify where the speaker grew up. When Gerrard took over the *Bookstall* series, he added his own flair for the dramatic. He didn't read books in a straightforward manner like Dumble. "I couldn't do it that way," he told an interviewer. "There's too much ham in me." Gerrard would go through books with markers and highlight the characters with different colors as cues to take on the proper persona and voice of each character.

Doing the show live "made me feel like I was on stage," Gerrard said. But live radio has its pitfalls. During one reading Gerrard had a bad coughing spell and pushed the "cough button" to switch his microphone off. He followed up with a few swear words and then released the button to continue. Before long, he saw an engineer holding a sign in the studio window: "Your cough button is broke." Such was the audience's love of Gerrard that there were no repercussions.

Gerrard tackled substantial books that took weeks of broadcasts to complete such as *Dr. Zhivago.* His reading of Orwell's *1984* was notable because a group of local mothers was attempting to ban the book from school libraries. He took listener requests and read all the books of popular hometown author James Thurber. In 1960, toward the end of his life, Thurber appeared in Columbus, playing himself in a revue called *A Thurber Carnival.* Gerrard arranged an interview for his show and went to meet Thurber at the Neil House hotel. He brought along WOSU's Don Davis to help with the recording equipment. As Davis told it: "We showed up right on time, and Thurber's wife came to the door, and she was in a robe and pajamas. This is 2 in the afternoon. She introduced us to her husband, . . . and he, like her, was in pajamas and a robe. And he greeted us and almost ran across the room to the telephone and picked it up. He started calling people all over Columbus; he had these phone numbers in his memory. Thurber was so excited because Gene Gerrard was here."

In the Bookstall remained part of the WOSU schedule for more than thirty years, ending in the early 1980s. Twenty years after that, the station was still receiving requests to replay some of the *Bookstall* programs.

CHAPTER 10

A CLASSICAL TRADITION

> Good morning, everyone, and happy Easter. This is Les Spencer opening the broadcasting day for WOSU and WOSU-FM, The Ohio State University stations in Columbus. Each Sunday at this time, your university stations present *Treasured Music,* a recorded program of serious music.
>
> —1956 RECORDING OF WOSU RADIO

When broadcast radio came into existence after World War I, more hours were devoted to music than to anything else. It was simply the best and cheapest way to fill airtime. At WOSU early staffers such as Ann Charles and Bob Higgy would often use their skills on the piano or violin to fill dead air. The station even created a volunteer orchestra. Listeners readily accepted musical programming, much of which was lengthy classical pieces. An argument can be made that radio saved classical music, or at least enhanced its ability to survive. In the 1920s the vast majority of Americans had never attended a symphony performance or an opera, but within a few years classical music had become a widespread form of entertainment on both commercial and educational radio.

Across the country, radio broadcasts of classical music led to a boom in community orchestras. One study found a remarkable increase in orchestras in small to midsized cities, from just sixty to two hundred eighty-six in the 1930s. Radio also propelled the sale of musical instruments; the sales of pianos during the Depression years exceeded those in any other period in history.

The sound on AM radio may have been "tinny" and lacking in clarity, but there was no basis for comparison and the sound improved with the introduction of the superheterodyne all-tube radio sets. Early radio was still strictly monaural, with everything channeled through one microphone and one speaker in the home, but it didn't matter. Music coming out of the big box in the living room was still magical. Classical music flourished when the NBC radio network committed to the genre and hired Walter Damrosch, who became one of the greatest classical radio music educators. Damrosch's mission was to upgrade the quality of musical culture in the country, and he had been at it for years before radio's existence. With this new means of reaching millions, he quit his job directing the New York Symphony to devote his career to becoming the musical impresario for NBC radio.

In Columbus the enthusiasm for classical radio was sparked by both the music and its many quality presenters. WOSU had a long history of talented music directors and classical announcers. They knew not only the music and the performances but also the stories behind them. They could say which recording of Wagner's *Tristan* was superior and talk fluently about the composer's love life.

This sophisticated on-air expertise started in 1939 with an unlikely source. Alfred Vivian's specialty was corn and oats, but his great love was Bach and Beethoven. Vivian led the growth of Ohio State's College of Agriculture until he retired as dean; he then became more well-known across Ohio for his weekly program on WOSU radio. *World Famous Music* sparked thousands of letters in its first two years of broadcasts. Vivian was the first WOSU radio host with a deep knowledge of classical music, and he was so popular that he hosted a second show, *Morning Melodies,* and a third on Sundays, *Treasured Music.* The music was drawn from Vivian's vast private library of classical records. Someone estimated they could listen to the great composers in his collection for thirty-seven days and nights without hearing the same selection twice.

One of the keys to Vivian's popularity may have been timing: Listeners found classical music to be a safe refuge from the often-upsetting news from Europe. Vivian died after just four years at the microphone, but WOSU's devotion to classical programming continued in no small part because he donated his immense library to the station. The windfall allowed the station to claim the largest classical collection of any noncommercial station in the country—twelve thousand records and five thousand electrical transcriptions.

Peter Allen grew up in Cleveland and served in the navy during World War II as a commanding officer of minesweeping vessels. After the war, he enrolled at Ohio State and became the principal violinist of the university symphony. As Allen tells it, one night he was invited to the WOSU radio studio to play in a string quartet, but none of the other musicians showed up. The station had to fill time, and Allen started talking to the radio audience about classical music and opera. Before he could leave the building, Allen had been hired as a part-time student announcer. Flash-forward three decades, and Peter Allen had become the primary announcer for Metropolitan Opera broadcasts in New York City. He continued as the voice of the Met for twenty-nine years.

In 1951 an OSU instructor in German, Glenn Goodman, mentioned to WOSU manager Bill Ewing that whoever was announcing the Friday night opera broadcasts was mumbling his words and had pronunciation issues. Ewing adroitly asked Goodman to step in until the station found a replacement. It took a decade for the replacement to show up as Goodman hosted opera and a variety of other classical programming on WOSU-FM until 1961.

FRED CALLAND

The same year Alfred Vivian died, an Ohio State student named Fred Calland was featured in a *Lantern* story about his own collection of rare recordings that numbered more than three thousand. Calland also collected the machines that could play his recordings, including hand-cranked gramophones and a nineteenth-century Edison phonograph that played cylinders.

Calland had served in Germany during World War II and then returned to Ohio State to further study the German language and education. In addition to teaching full-time, he helped with the German courses on the Ohio School of the Air with Ilsedore Edse.

Calland was asked to take some on-air music shifts at WOSU, and his radio personality, sense of humor, and musical knowledge caught the attention of the listeners and station managers. In 1955 he became the station's music director. Verena Dambrans, a native of Latvia, volunteered at WOSU to help classical announcers with their pronunciation of the names of Russian composers, conductors, and musicians. Dambrans described Calland as a tall, handsome, mustachioed gentleman and the most knowledgeable person about classical and operatic music she ever knew. Despite a penchant for practical jokes and a biting sense of humor, Calland had strong views about classical music and opera. "Only serious music is played at WOSU—music that is intended to be listened to with full attention," he said.

STEREO INNOVATION

After WOSU-FM radio went on the air in December 1949, most of the same programming that had aired on the AM station was simply simulcast on FM. After all, FM radios were still few and far between. However, because the AM station was a daytime only station, classical music soon crept into the evening hours on FM. By the mid-1950s the creative WOSU engineering team had the notion to experiment with the two stations to coax more listeners to buy FM radios. Announcers asked listeners to set up two radios and tune in to 820 AM on one and 89.7 FM on the other. At the start of every such "stereo" broadcast, the announcer would tell listeners which station had the right channel and which had the left.

The innovation proved popular, and the station aired a thirteen-week series of stereophonic broadcasts every Sunday afternoon. Donald Lakin recalled experiencing WOSU's version of stereo radio as a twenty-year-old: "My dad and I, much to my mother's chagrin, set up speakers in the corners of our living room. We had two tuners and amplifiers upstairs with wires that passed through our house on Livingston Avenue. I think I listened to every stereo broadcast the station produced." In a quirky 1957 television promo for stereophonic radio, Fred Calland, in a fake goatee, directed singers John and Mary Schmidt, one representing FM and the other AM, while he explained the difference between monaural and stereo radio. It was a great glimpse of some of the fun the staff had with this new medium called television.

It wasn't until the early 1970s, when WOSU-FM replaced its transmitter, that real stereo broadcasts on FM became possible, but for many years WOSU took great pride in providing the only stereo radio broadcasts in Columbus.

WOSU's classical music franchise forced the Cincinnati community to develop its own. In the late 1950s, Cincinnati radio stations aired only four hours of classical music a week. A group of local leaders saw that WOSU was airing some forty-eight hours of classical and opera music every week and started working with station management to bring an FM radio station to Cincinnati that would repeat the WOSU signal. The idea of a Columbus station invading the community wasn't popular, and soon the University of Cincinnati acquired a noncommercial license. WGUC went on the air in 1960 to serve the pent-up classical needs of the Queen City.

Some of the first off-campus program support came to WOSU due to classical music. In 1964 WBNS radio dropped their affiliation with the New York Philharmonic, and WOSU's radio program director Ken Keller decided to go after the contract. He forged an agreement with the research giant Battelle Memorial Institute, located next to the OSU campus, which covered the $2,900 for the telephone circuit to New York in exchange for on-air mentions of their cutting-edge research.

AT THE OPERA

The tradition of broadcasting opera had started in Chicago in late 1921 when station KYW set up its microphones to capture the entire schedule of the city's Civic Opera Company. In 1931, the powerful clear channel commercial station WLW in Cincinnati was carrying the Metropolitan Opera from New York City every Saturday afternoon and booming its signal across the country. During World War II the NBC network and WCOL in Columbus picked up the Met broadcasts. By the mid-1950s the Met was airing on educational radio stations that had newly established FM stations. That was the case with the regular live broadcasts that started more than sixty years ago on WOSU-FM. The Met broadcasts continue today on WOSU's Classical 101.1 FM.

CHAPTER 11

SAVING EDUCATIONAL TELEVISION

Educational television must not be given
the giblets of the television turkey.

—FRIEDA HENNOCK, FCC COMMISSIONER, 1951

In the summer of 1948, during a trek across Ohio to visit the commercial television stations that had just gone on the air, Paul Wagner and Harrison Summers found little confidence in the medium. The two members of the university Radio Advisory Council reported that local station owners were not optimistic, saying television was a "costly and uncertain venture at best" and that the quality of the picture was lacking. Audiences were also small to nonexistent in Ohio. While about one hundred thousand TV sets were in use in the United States that summer, two-thirds were in the New York City area, and most of the rest were in Chicago and other large cities. Midwesterners could read about television, but most could watch it only in department store windows.

Following the trip, Summers reported that it would be nearly impossible for Ohio State to establish its own station, because the cost of doing so would be "almost four times the yearly budget of WOSU radio." He suggested that the best way for the university to get involved in television was to work with local commercial stations as they came on the air to provide training opportunities for students. A year later WLW-C (NBC), WBNS (CBS), and WTVN (ABC) became the area's first television stations. Collaboration with Ohio State began shortly thereafter through coursework and student volunteers. For institutions such as Ohio State, which had invested in AM radio for decades and more recently in a new FM-licensed station, the thought of coming up with more funds for a TV broadcast station seemed improbable.

The reality in the late 1940s was that without an FCC reservation of channels for education, few institutions would risk taking the plunge into television. By 1949 just five universities were seriously involved in planning for television, according to a short history written by Richard Hull, who later became director of WOSU. Ohio State was not among them. The station Hull helped put on the air at Iowa State University, WOI-TV, was the first nonexperimental, educationally owned station in the country. It was considered a commercial operation because that was the only designation available.

When the Federal Communications Commission held preliminary hearings on channel allocations in 1948, few educators appeared at the hearings, and the National Association of Educational Broadcasters (NAEB) noted with alarm that "once the TV channels are filled there is likely to be no second chance" for educational television. At Ohio State's Institute for Education by Radio that year, the hot topic was television. C. M. Jansky, a leading engineer at WHA in Wisconsin, proposed a plan for educational reservations in the ultra high frequency (UHF) spectrum, which was unoccupied territory, to serve as a "hedge" until educators came around to television.

The NAEB subsequently adopted a resolution urging the FCC to reserve some television channels for education. Still, the majority of the educational radio license holders were indifferent. The rash of commercial applications for early television licenses mirrored what had occurred in radio and caused similar chaos. This time, the FCC decided to freeze all pending and future applications and work on allocating licenses community by community. The freeze was expected to last one year but stretched into four after the outbreak of the Korean War. It was a stroke of luck for educational broadcasters. The delay allowed the true believers, who strongly advocated for the use of television for educational and noncommercial purposes, to gain additional advocates and organize.

The initial allocation table developed by the FCC didn't reserve a single channel for education, prompting outspoken commissioner Frieda Hennock to file a lone but strong dissent. If there was going to be any educational television, this was the moment to act.

Dick Hull at Iowa State defined the place and time where the educational television movement took root: the Allerton House at the University of Illinois in 1949. Funded by the Rockefeller Foundation, educational radio broadcasters met to devote two weeks to discussing the potential and the challenges of getting involved in television broadcasting. There were intense philosophical discussions as well as debates about practical considerations, including the need to create a national television network of programming and a well-financed production center. But if the FCC did not set aside channels outside the commercial realm, none of it would be possible. The Illinois conference remains "the philosophical cornerstone of public broadcasting," according to some historians.

Coming out of the conference, educators knew they needed a coordinated voice to represent their interests in Washington. Ohio State's Keith Tyler suggested an organization of groups knit together by common goals and leadership, and the Joint Committee on Educational Television (JCET) was created, with Tyler chosen as its chair. The organization's one mission was to secure channels from the FCC for educational television. Such an effort had failed for AM radio, but the more recent success with FM gave the JCET leadership hope.

Tyler thought the effort came just in the nick of time. Educators "had been dragging their feet," assuming that success with FM radio allocations would automatically transfer to television. Instead, egged on by Commissioner Hennock, they needed to make a compelling case in Washington.

THE FCC HEARINGS

On October 20, 1950, Ohio State president Howard Bevis received a special request from Dick Hull, then serving as president of the National Association of Educational Broadcasters. He asked that Ohio State allow Keith Tyler to take an extended leave of absence to organize the campaign to reserve educational frequencies. With that approval, Tyler moved into space in Washington provided by the National Education Association, a former bathroom at the old Guggenheim mansion that began to be called "Keith's powder room." The JCET sought funds to support its efforts from various foundations and universities, writing bluntly that "if we do not act now, this magnificent educational and cultural medium will be lost to the people irrevocably."

A series of JCET sessions filled with heated discussions ultimately brought together what seemed like a strong case. Hull advanced a concept to allow limited advertising to cover some of the operating costs of educational stations, but this drew opposition, with most of the group believing pressure to sell ads would distort educational objectives. The JCET wanted to stress VHF reservations over UHF, since television sets made at the time couldn't even receive UHF signals, but that strategy would change.

The single advocate for educational television on the FCC was Frieda Hennock, and the group sought her advice. Clad in a fashionable hat and mink stole, Hennock spoke to hundreds of educators at Tyler's radio institute in downtown Columbus in 1950. "Educators need to get organized to do the job," she said. "If they don't fight now they will have to take the crumbs from commercial TV, whose primary interest is entertainment."

The 1950–51 FCC hearings were seen as the last chance to reserve noncommercial television channels for education. As the hearings began, Bevis testified in his role as chair of the Land Grant College Association's Radio Education Committee. A basic tenet the JCET advanced through Bevis was that the TV spectrum was a natural resource owned by the public, just like national parks. Bevis also believed the same principles that applied to land-grant universities like Ohio State, which received federal lands to establish public universities, should apply to broadcasters seeking a portion of the spectrum for public use.

The lack of programming emphasizing children and local culture was seen by the JCET as a great deficiency in commercial broadcasting. Harold McCarty of Wisconsin's WHA emphasized the FCC's responsibility to the public to fight off the pressures of commercialism and recognize a higher calling. "If we are to have any substantial educational development in television," McCarty said, "it will take place only if education has a place of its own."

Testimony from educators revealed that though there was great interest in developing television on university campuses, few institutions would be able to activate channels immediately. Hennock told her FCC colleagues: "Let us say you reserve them and nothing is done by the educators. . . . Still nothing is lost. This will give these educators an opportunity to get in there and pitch." Hennock's philosophy became a valid rationale for many of the commissioners. She was such a strong presence that commercial television's *Broadcasting Magazine* complimented her in its December 4, 1950, issue: "Commercial broadcasting, the Commission itself and the FCC staff were virtually taken to the woodshed as Commissioner Frieda Hennock championed the educators' cause. Hennock is quarterbacking the educational strategy . . . and has been making touchdowns in every quarter."

In his testimony, Bevis emphasized that if all of television went to commercial interests, it would be "unthinkable both in terms of the historic role which education has assumed in our society and the record of AM broadcasting where no special reservation was made for education." He touted the current work of WOSU radio: "WOSU directs its programs to farmers, to homemakers, to parents, to boys and girls in Ohio's classrooms, and to interested citizens generally. WOSU-FM has enabled the University to extend its broadcasting into evening hours. Now we are alert to the great possibilities of television. Some of our departments are already experimenting with educational series on local commercial stations, acquiring experience and 'know-how' in this new medium."

Among the many who testified during the hearings were two prominent Ohio politicians. Congressman John Vorys spoke about how important it was for communities such as Columbus to have an educational channel, but Ohio senator John Bricker was Keith Tyler's secret weapon. With his conservative Republican values, Tyler said, one would expect Bricker to be opposed to "turning the public domain over to education." Instead, Bricker testified with force that educational set-asides for public service programming were vital for the continued freedom of the country.

At one point, an FCC commissioner asked about noteworthy examples of educational radio excellence. A commercial broadcaster cited WOSU and WHA as "two exceptions to the dismal failure" that was educational radio, insinuating that such a failure would repeat itself. While that notion was quickly debunked by the JCET lawyer, the comment elevated the status of the Columbus and Madison stations.

One notable witness for the commercial interests was Frank Stanton, the president of CBS, who had earned his advanced degrees at Ohio State. Stanton said the FCC's priority should be establishing a strong commercial television industry to create a mass audience, which TV had started to do with "the Arthur Godfreys and the Milton Berles." Stanton believed educational reservations would be a waste of spectrum because most educators simply would not use the channels.

Some of the commercial networks had been expecting an easy fight to reserve all television channels for their use, calling anything else "illogical." They were surprised by the thousands of pages of testimony by educators and supporters coming out of the initial FCC hearings. In *Broadcasting Magazine*, Larry Christopher wrote: "The vigor of their attack indicated educators did not intend to be left out in the shuffle as they were in the history of AM radio."

One of the strongest arguments by educators came late in the hearings when the results of a recent study of commercial television broadcasts in New York City were revealed. It showed that only a small portion of the broadcast schedule was used for children's programming or cultural, instructional, or news programming. What was missing on television in the country's largest metropolitan center was programming about the arts; news and public affairs; in-school programs; special education programs; adult courses; and more. "Commercial broadcasting was stubbing its toe all over the place," said Tyler. "People were getting concerned about the tripe of broadcasting. They were getting disgusted with the programs. They were repetitious."

THE DECISION

The most important decision in public television history occurred on April 11, 1952, when the FCC released its Sixth Report and Order, outlining a new allocations table for television with 2,053 assignments in 1,291 communities. Noncommercial educational television received 242 reservations: 80 on VHF and 162 on UHF.

As word of the decision trickled out in Washington and beyond, the JCET group of twenty leading educational broadcasters from across the country was triumphant. "When we won the reservations, there was a tremendous feeling of jubilation and victory at the JCET offices. . . . It was like an election headquarters looks after a victory," said Dick Hull. Keith Tyler added: "Everyone had been

FIG. 11.1. Several of the primary advocates for the reservation of educational television channels celebrate their success at Ohio State's Radio and TV Institute in 1952. Note the educational TV allocation map in the background. From left to right are Ralph Steetle and Keith Tyler of the Joint Committee on Educational Television, FCC commissioner Frieda Hennock, and FCC chairman Paul Walker.

working literally day and night in our efforts to secure noncommercial reservations. . . . We felt we really deserved and had earned the favorable ruling. . . . The win brought a feeling of great satisfaction." Tyler celebrated that afternoon with FCC commissioner Frieda Hennock and others at the Raleigh Hotel in Washington.

The FCC decision is considered a landmark in the history of broadcasting and education in the United States in that it reserved both natural resources in the form of television spectrum and specific channels for use by educational and nonprofit institutions. The group that led the charge believed that the potential of television was limited only by one's imagination. Ohio State's Keith Tyler had provided a steady but urgent presence to the process. And among the many who testified, the Ohioans—Bevis, Vorys, and Bricker—were highly effective.

Hartford Gunn, the first president of the Public Broadcasting System, spoke about the Joint Committee on Educational Television and its accomplishment at the PBS annual meeting in 1979: "Thirty years ago, a few far-sighted people—educators, mostly—saw a potential for using this new technology as a way of reaching people with a service that had been limited by what a textbook and chalkboard could do. Had these pioneers not seized the opportunity, if they had failed to see and take advantage of the window of opportunity—we would not be here debating the future of the system today. There wouldn't *be* any system; there wouldn't be any public television at all."

CHAPTER 12

UHF 34

> This is a very significant day in the history of The Ohio State University. The beginning may be small and may attract little public attention. But from this seed which is planted today there will surely come growth and productivity beyond our dreams.
>
> —DR. FREDERIC HEIMBERGER, OHIO STATE VICE PRESIDENT

With the great success in Washington at setting aside channels for educational television, WOSU leadership expected the university to move quickly to establish one of the early educational television operations. Bob Higgy developed the technical plans, and Keith Tyler set about creating the case for WOSU to gain an allocation of VHF channel 12 in Columbus. Given the station's status in educational broadcasting (and Tyler's status in leading the JCET), Ohio State thought it had a shot at the coveted VHF frequency. WOSU was desperate not to be saddled with the only UHF frequency in a town already full of VHF commercial stations.

The proposal showed that with some adjustments to the allocation of a channel in West Virginia, the university could be granted channel 12 without any fear of interference issues. The FCC denied the request. Although the allocation table had attempted to reserve educational VHF channels for major cities, it wasn't possible in Ohio. The band was already full of commercial broadcasters in Cleveland, Cincinnati, and Columbus by the time the FCC set aside channels for educators. The plan shifted to seeking channel 34, and the proposal landed on the Ohio State trustees' agenda in 1952.

The problems with UHF were plentiful, as outlined to the board by Ohio State president Howard Bevis:

- There were no UHF transmitters available for purchase, although manufacturers promised development within eighteen months.
- There were no television sets that could receive UHF in homes, although they were being built and adapters were becoming available.
- A WOSU station on channel 34 would be stranded on the UHF band without any benefit of commercial stations bringing audiences to the higher band.
- The high expense was an additional consideration, including the need for a new building to house WOSU-TV studios.

Despite these and other challenges, the seven trustees, including Senator John Bricker, unanimously recommended that the university accept the channel and begin to establish the educational television station. Tyler, Higgy, and the WOSU staff were gratified by the vote of confidence and quickly went to work. Tyler sought support for a new position to coordinate the Ohio State programming then airing on local commercial TV and start experimenting with program concepts in partnership with university departments. He thought it was important to expand on this production experience to develop know-how in the use of television in education.

Bevis and the university administration advocated a much slower pace. They saw Tyler's proposal as too much, too soon and refused to support the new position. Tyler fired back, saying the decision would "seriously cripple progress toward the operation of the station." He pointed to the foresight of leaders at Michigan State and Wisconsin, who built on-campus studio space and hired staff before actually going on the air. He also said the lack of support was a "serious blow to the University's leadership and prestige in the field."

Within a year the FCC granted channel 34 for use as an educational television station at Ohio State, but there was still no plan for investing in the development of the station. University leaders were finally spurred to action by an offer from the Ford Foundation to provide a matching grant of $100,000 for capital costs to build the station. It was one of numerous grants to educational television start-ups across the country.

A new WOSU-TV building committee was formed, directed by Frederic Heimberger, vice president for academic affairs. The committee had trouble finding space for a TV studio on the central part of campus, so focus shifted to the western edge, on farm acreage used by the School of Agriculture. The idea of moving the radio operation into the same building was considered but quickly abandoned because of the expense. Radio would stay where it had been since 1920.

Despite his national prominence, university leadership started to bypass Keith Tyler. He was left out of much of the television planning process. Tyler showed his disappointment by sending a lengthy analysis to Bevis that reviewed the history of educational television and the current "sorry state" of WOSU television's development and supervision.

Reacting to Tyler's criticisms, including the "confusion and lack of direction" in the planning for WOSU-TV, the Heimberger committee advised Bevis to hire a director for all of Ohio State's broadcast properties, radio and television, at the level of a major college dean. Such a hire would be delayed until after the TV station was actually built.

Construction finally began in November 1954. The unassuming single-story, yellow cement structure on North Star Road was built for a final cost of $123,257, and by the summer of 1955, television equipment was arriving at the loading dock. Chief radio engineer Harold Gorsuch and other technical staff handled the integration of all the new toys under Bob Higgy's supervision. Higgy may not have been the most effective leader during his decades at WOSU, but his engineers were in awe of him. "Once in a great while the spotlight shines on someone who has the necessary self-confidence, the necessary wisdom and determination, and who also understands the laws of physics," wrote WOSU engineer Edgar Lee Howard.

The youngest person on the installation team may have been Robert Butche, a student with the rare distinction of having an FCC First Class Radiotelephone Operator License. He knew WOSU well from having performed in the radio station's theatre productions as a youngster in the University School. Butche found the work hard, hot, and memorable: "None of us had ever seen any of this equipment before. When we finally turned on the transmitter for the first time and it didn't melt and the feed line didn't melt, we knew we were radiating a signal, and we thought, by golly, this is impossible because this was at a frequency above what anybody could've imagined. This was up in Ultra High Frequency. So it was a big deal. I don't think anybody could receive it, but we took a receiver down to the Statehouse and got a signal. That was important for political reasons."

The station sent out regular transmission tests with films from Ohio State's photography department. Members of WOSU's radio production team had seen but never touched a television camera, so they headed west to a TV workshop at Iowa State.

All the delays in getting the television operation approved and built meant Ohio State had squandered its chance to be among the first educational stations to hit the air. In 1953, KUHT at the University of Houston became the first to broadcast in the new category of noncommercial educational television. A year later WCET in Cincinnati was the sixth educational station in the country and the first in Ohio. To the frustration of broadcast pioneers Higgy and Tyler, WKAR at Michigan State, WHA at the University of Wisconsin, and WILL at the University of Illinois all beat Ohio State to the air. What Tyler had warned Bevis about had come true: Ohio State's leadership in the field of educational broadcasting had started to fade.

THE PREMIERE

Sitting at a desk (scavenged from surplus property) in front of a large microphone, the telegenic John Schmidt spoke slowly to the camera, announcing the coming of a new television station to Columbus. The surviving film shows Schmidt glancing at a script as he described the upcoming schedule of the station, which was yet to hit the air. Schmidt outlined the program schedule of four hours a day during the week and two hours on Saturday nights for a total of twenty-two hours weekly. Whether the film aired on local commercial television or in movie theatres remains a mystery, but it was the first of many promotional spots touting WOSU-TV.

On February 20, 1956, the cover of *Time* magazine featured an artist's rendering of Frank Lausche, who was in his fifth term as Ohio's governor. The story was about the Democratic leader who talked and acted like a Republican, remaining popular even though "big-shot Republicans resent him; organization Democrats detest him; labor leaders denounce him; and his immigrant parentage arouses suspicion." It was the talk of Ohio.

The team at WOSU was otherwise occupied that Monday. At precisely 3:40 p.m., after airing a film about the Ohio State campus, Schmidt faced the massive scoop lights flooding the jam-packed studio. He ad-libbed an official first greeting to the people of Columbus and then turned the microphone over to Frederic Heimberger, whose committee had advanced the plans for the new WOSU-TV building. "This is a day of great significance," Heimberger said. "With full knowledge of the difficulties involved, this university has been happy to seize this new opportunity to accept this challenge. It is our firm conviction that, with proper development, this new medium has possibilities beyond the bounds of imagination."

Heimberger went on to describe how history was repeating itself. He pointed to Ohio State's founding as a land-grant university and its charge to bring the benefits of education and research to the citizens of Ohio and beyond. "Now again, we have been given an opportunity and a challenge to extend the services of this University to a vastly larger audience. The goal is to make the most of educational television and radio as means for extending to the people of Ohio the best that can be offered."

It may have been a cool 44 degrees outside the studio, but inside the heat was becoming unbearable. Behind Heimberger, the glaring klieg lights showed an impatient sixty-voice symphonic student choir directed by Louis Diercks. When they finally got their moment, the students stepped up and delivered quality renderings of "Let Thy Holy Presence," "How Long, O Jehovah?" and "Been in the Storm So Long," which may have been what Bob Higgy was thinking as he watched the proceedings. At 4:06 p.m. the local programming was over and WOSU started airing a series of national film documentaries. By 8:30 it was showing a slide with a test pattern and a tone; then the loud humming of the transmitter tubes faded away as the station was shut down. The staff was jubilant and relieved. The only glitch had come between program segments when someone put the "One Minute Please" slide on the screen upside down.

Rough estimates had the number of potential Columbus-area viewers able to watch the first broadcast at two to three thousand, less than 1 percent of the total number of televisions within the

FIG. 12.1. WOSU-TV's first broadcast from the studios on North Star Road may have been seen by a few hundred people. Announcer John Schmidt introduced OSU vice president Frederic Heimberger to unveil the only UHF station in Columbus.

45-mile coverage area of channel 34. Actual viewership was likely a few hundred, although no one really knew. One woman who saw the broadcast called WOSU to learn more. When told it was an educational venture, she said, "No commercials? You'll be out of business in a month!"

More representative was the letter to the *Dispatch* from a woman who didn't believe that another station was on the air at channel 34 because her TV dial only went up to 13. But some who saw the first days of programming found the station to already be a beacon of learning. Freda Koch, who wrote a column for the *Clintonville Booster,* a suburban Columbus weekly, had become one of the station's first fans:

> I'll tangle with anyone who says that educational television will have little to offer. Educational television has something new to offer. I spent five minutes with John Schmidt giving headline sports.

(Handsome man, this Schmidt. He lives on Brevoort Road.) Then I went to the jungles, a logging mill, and heard a captivating history of the royalty of wood—mahogany.

My travels then changed to a trip to Honduras. No ordinary film this. It told me how Christopher Columbus set foot there first in his journey to the new world. I had been traveling all evening, right on my own davenport. And during all this time I had not had one program interrupted by a commercial. I had been (perish the phrase) glued to my seat and I had learned something.

My reaction at the time the station signed off was that educational TV was bringing a kind of tranquility to television. I was grateful to my state of Ohio. I do thank it and I'll keep watching. Can't resist."

WOSU was the nineteenth educational television station on the air in the country. Despite FCC warnings to develop stations quickly to assure that educational frequencies remained available, most assignments were unfilled until an institution or a community nonprofit picked up the challenge. By 1960 only forty-nine educational television stations were on the air, but the two-hundred-plus allocations left were still of great interest to many educators and communities. Cleveland's channel 25 allocation was filled in 1965, when WVIZ, a community nonprofit licensee, became the one-hundredth noncommercial station.

At WOSU the task for years to come was to educate viewers about how to find the only UHF station on the dial. The engineering team sketched out a basic guide that was distributed to anyone who wanted one: "Make Your Set an All-Channel Receiver—It's Easy—In 10 minutes." The guide described the UHF band of channels 14 through 83 and how to receive channel 34 on a designated VHF channel using a special UHF antenna and converter. It concluded with an option to call WOSU-TV and ask the engineering department to troubleshoot.

Michael Schmidt (no relation to WOSU announcer John Schmidt) recalled his father's purchase of a Blonder Tongue brand UHF converter when he was a young boy in Columbus. "The converter was attached to a black-and-white TV, one with a Bakelite body and a screen of about eighteen inches," he said. "WOSU was the only UHF station at the time, and the converter got a lot of use

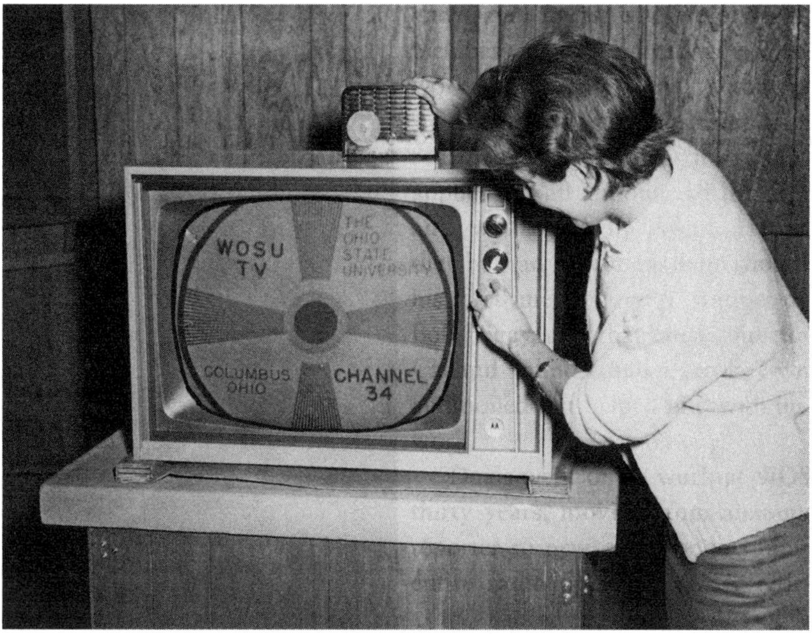

FIG 12.2. Station engineers installed and serviced UHF converters at the request of WOSU-TV viewers, making more than fifteen hundred service calls in the first year on-air.

and certainly was worth the investment. We all grew up as WOSU fans."

By 1962 only 5 percent of televisions in the country were able to receive a UHF channel without a special converter. However, the All Channel Receiver Act, which was passed that year, required new sets to have both UHF and VHF tuners within two years. More than 42 percent of televisions could receive UHF and VHF by 1967, which helped stations such as WOSU begin to find a larger audience.

DICK HULL TAKES OVER

At Iowa State, Dick Hull's work in early television was groundbreaking. He also served as president of the National Association of Educational Broadcasters and worked closely with Keith Tyler to help assure the reservation of educational frequencies. Hull knew Ohio State and Columbus well. As a student at Iowa State in

1934, he had participated in an exchange program and worked as an announcer for Higgy on WOSU radio for six months. He later attended numerous gatherings of the Institute for Education by Radio, and in 1955 he received one of the institute's twenty-fifth anniversary awards for his leadership in developing WOI-TV.

Hull was busy wrapping up a cross-country evaluation of educational television as part of a foundation grant when Ohio State approached him about the job of directing all its broadcast properties. He countered with a list of demands that included a full professorship, a reporting line directly to the president's office, and adequate funding to properly grow the staff, equipment, and facilities.

In June 1956 the forty-two-year-old Hull was introduced as Ohio State's new director of radio-TV broadcasting. He was WOSU's second full-time director, with Higgy having served in that role since 1927. His $18,000 starting pay was nearly double Higgy's current $10,000 salary.

Hull immediately took over, delegating all the day-to-day operations of radio and television to Bill Ewing. Though he named Higgy to run all engineering, within a year he relegated Higgy to a special-consultant role and liaison with the FCC. Al Boggioni and Harold Gorsuch would engineer the radio and TV facilities. David Ayers moved from radio to become the programming and production manager of the TV facility. Radio personalities Fred Calland and Gene Gerrard, newsman Don Davis, and radio programming director Don Quayle were regularly called upon for television production and on-air work. Radio was now the old technology and relegated to the background. Hull found a trusted assistant, Martha Haueisen, who directed business operations and gained enormous influence. Ayers remembered her as a strong and feared gatekeeper: "If you crossed her, she would take your keys away." Engineer Edgar Lee Howard called her Hull's "personal secretary, office manager, bookkeeper, and accountant who truly ran the stations day-to-day, while Mr. Hull was busy being a visionary, consulting, and making speeches."

As the station's chief announcer, John Schmidt also produced many early shows, including the first local children's show, *Kid's Stuff,* and a popular music series, *Songs for a Summer Evening,*

FIG. 12.3. A rare photo of WOSU's leadership in 1957 including (left to right) Robert Higgy, Richard Hull, and William Ewing.

which featured John and his equally attractive wife, Mary, singing Broadway selections. Schmidt had grown up in Columbus and was an All-American pole vaulter at Ohio State. He announced and sang on WOSU radio before finding a career on Broadway, where he met his wife. In a 2017 interview, Schmidt said the numerous live pro-

FIG. 12.4. John and Mary Schmidt hosting and singing for *Songs for a Summer Evening,* a live musical series that began in July 1956.

ductions at "WOO-SUE," as he called the station, were "very exciting, anything could happen, but you got used to it. They wanted a talker and I could do that."

Schmidt once hosted a series called *Relax and Live* that included a program on gun safety. He recalled picking up a shotgun during the show and taking aim at the camera in fun. The instructor with Schmidt gingerly took the gun out of his hand and later explained to the shocked host that it was loaded and his finger had been on the trigger.

Shortly after Hull was hired, OSU President Howard Bevis retired and was replaced by Novice Fawcett, a former superintendent of the Columbus City School system. Fawcett had supported the school system's efforts to put its station, WCBE-FM, on the air in 1956 and believed in using radio as an educational resource.

EARLY PROGRAMMING

Programming on WOSU-TV in the early years included films provided by the National Educational Television network, including

Frontiers in Space, French through TV, Patterns of Life, and *The Great Plains Trilogy.* The NET provided films and later tapes to more than one hundred educational television stations. Each week, WOSU could expect to receive five hours of new NET programming, and it broadcast all of it.

Dating back to the early radio days, WOSU had built connectivity to campus buildings to expand its programming capabilities. It did the same with television. Using new image orthicon tube cameras, with giant pedestals, along with controls, cables, monitors, and power generators, every setup required extensive time to complete. The equipment was stacked into a used twenty-five-passenger bus that had been donated to the station and taken to a variety of campus locations for live programming. TV managers crowed about how the entire Ohio State campus was becoming one vast studio. The March 1958 staff newsletter reflected the ambition of the production team: "*Camera on Health* will originate from the surgery wing of University Hospital tomorrow. Then the camera equipment will be moved to St. John Arena for the Winter Commencement followed by the State High School Basketball tourney over the weekend. Rounding out the week is the address of the US Secretary of Agriculture Ezra Taft Benson at the Ohio Union."

Local programs included the Ohio State Award–winning series *German Hour,* which had a following carried over from its many years on radio. John Morison produced *The Seventeenth Star,* a series on Ohio history. Schmidt and Keith Tyler hosted *Diary of a Traveler,* interviewing prominent visitors to campus. There was *Mr. Lincoln,* a biography based on writings of Lincoln authorities, and *Research Doctor,* a series on medical research at Ohio State. *Essentials of Freedom* was based on a symposium held at Kenyon College. It was produced and distributed by the Department of Photography to a network of fifty-two educational stations and won the George Washington Honor Medal.

A new children's show called *Five and Ten* became popular after premiering in the fall of 1956. Aimed at youngsters between five and ten years old, it built on the concept of a variety show with games and stories and puppet characters, not unlike the national phenomenon *Sesame Street,* still a decade away. The host would introduce a game to play with the viewing kids and include dances and stories to keep them engaged and moving.

FIG. 12.5. Peggy Bugg hosted *Five and Ten,* a popular children's show on WOSU-TV that included two muppet mice.

Dick Murgatroyd produced the series and expanded the show in its third season to a daily 5 p.m. broadcast. Recurring characters included two sibling puppets, Mr. and Miss Mouse, who would talk about the problems of being a mouse. Tom Patterson, aka "Mr. Magic," would astound viewers with sleight of hand, and David Jones of Ohio State's veterinary medicine school would introduce a variety of animals. Peggy Bugg, a kindergarten teacher who was as cute as her name implied, hosted the show during much of its run. *Five and Ten* was such a hit that it was later fed by microwave to other stations in Ohio.

WOSU-TV was producing an enormous load of creative programming. "We did some imaginative things despite the many limitations we had," recalled David Ayers. By 1961 WOSU was on the air from 8 a.m. to 10:35 p.m. Monday through Friday, but remained off the air most weekends.

THE WONDER OF VIDEOTAPE

"It's difficult for many to realize the true significance of this gift," said WOSU's Harold Gorsuch; "the possibilities are fantastic." As the television operation's technical supervisor, Gorsuch was very excited to install the first videotape recorder in Columbus. A new invention from the Ampex Corporation, videotape quickly revolutionized television. WOSU received the recorder in 1959 through a grant from the Ford Foundation. It was a massive two-inch reel-to-reel tape machine that could capture studio productions and instantly play them back, with no expensive film processing required. As program director, David Ayers was excited at the opportunity to tape programs and events to show at more convenient times on TV. He also put hosts and teachers in front of the machine so they could review their on-air work and improve their presentations.

Many videotaped studio programs were soon produced in collaboration with university departments, including *Opera House* with Theron McClure of the School of Music; *In Rehearsal,* which went behind the scenes of student performances; and *Aspects of Art* with Hoyt Sherman of the School of Fine and Applied Arts.

For the first time, sports programming such as basketball, wrestling, track, and swimming meets could be taped for later broadcast. WOSU television and radio both broadcast all home basketball games during the magical season of 1959–60, when the Buckeyes with star players Jerry Lucas and John Havlicek won their only national basketball championship.

With little available in the way of network programming, the North Star studios were constantly abuzz with local efforts. The station was so crowded that the men's room had to be used as an office—while also remaining a men's room. A five-room frame house was built adjacent to the station to ease the crowded conditions.

By 1959, WOSU had acquired a donated passenger bus nicknamed "The Wire." The bus was repainted to show off the WOSU brand and retrofitted to carry remote equipment, a video switcher, and one of WOSU's new tape machines to Ohio State athletic events. The used bus quickly developed a bad reputation among staff. Ayers and Schmidt recall a 1961 road trip to tape a football game at Northwestern. The bus broke down short of the Northwestern campus, and they had to push it to the stadium while Wildcats fans cheered them on.

Ken Keller remembers doing color commentary for an OSU basketball broadcast at St. John Arena. The broadcasts out of the remote bus were often produced with just two cameras, already offering minimal coverage of a game, but Keller noticed on his monitor that only one camera seemed to be functioning. He called down to the bus to talk with producer Dick Murgatroyd, saying: "You do know you've only got one active camera, don't you?" As Keller writes, "Dick let out a melancholy sigh and replied, 'I do, and we hope to do something about it, just as soon as someone puts out the fire.' A goodly portion of the inside of the bus was in flames."

TELEVISION COURSES

It is hard to fathom, but in 1916 the University of Iowa had a faculty member sending out via Morse code a series of lectures to amateur radio operators across the state. It was one of the earliest technological efforts to extend the benefits of a university to citizens unable to attend classes on campus.

Such initiative reached new heights with the Ohio School of the Air on WOSU radio and the Radio Junior College during the Depression years. With the coming of television, it was time to experiment all over again with using the medium for formal education.

"All you need do is sit in on a zoology class taught by Ohio State's new closed circuit television network to gain an awareness of the changing role of the University in the 1960s," wrote Novice Fawcett in his president's report for the first year of the decade. College administrators across the country were bracing for the flood of early baby boomers born after World War II who would soon invade campuses.

The development of televised courses was part of the answer promoted by Dick Hull with a team that included his chief assistant, Bill Ewing, and producers John Morison, Robert Fuzy, and Bill Buccalo. Over the next decade the courses became a profitable revenue stream for WOSU.

The first TV class in 1958 was a required college-level remedial course in math. Such lessons were a constant concern because as many as one-third of Ohio State students needed the instruction. In 1961 televised instruction accounted for 52 percent of all airtime on WOSU-TV, or more than seven hours daily. The following year, over seventeen thousand students were enrolled in TV courses through WOSU facilities, including two thousand students taking Zoology 400.

Student surveys soon identified complaints ranging from poor audio quality to lights that could not be dimmed during classroom telecasts. Despite the concerns, televised instruction continued to grow as students poured through the university's doors. By 1965 WOSU was using television to teach sixteen introductory courses reaching twenty-three thousand students. A closed-circuit television system now linked thirty-one classrooms in eight buildings. Custom studios were created on the top floor of Derby Hall to handle the productions and help alleviate the pressure on WOSU to produce and broadcast the courses.

Complaints continued. John Morison found some students had difficulty paying attention when the focal point was a 23-inch TV screen. "There is a constant fight to keep students' interest," said Morison. One student commented in a survey that "all TV instructors should pass a course in public speaking."

However, some students loved aspects of the teaching, such as the one who saw a cell nucleus on TV in zoology class. "It was perfect," she said. "I never saw one that clearly, even with a microscope in the lab." In fact, students generally gave televised science classes higher ratings than others. The telecourses were most effective when the video feed supplemented teachers in the classroom, using television to its greatest advantage. Under a grant, the College of Medicine worked with WOSU to allow video to be sent from operating rooms and research labs directly to classrooms.

"Dick Hull had quite an influence on using video in teaching and training," Morison recalled. "At the Dental School he worked with the faculty to change the way they taught operative dentistry." Morison described putting a camera against a patient's mouth so some seventy-five students could see in. "The cameras were large and the lighting was tricky. I thought I knew what a close-up was until we filled the screen with a tooth in the rear of a mouth."

By the early 1970s complaints from both students and faculty resulted in cutbacks in television course production, and the need

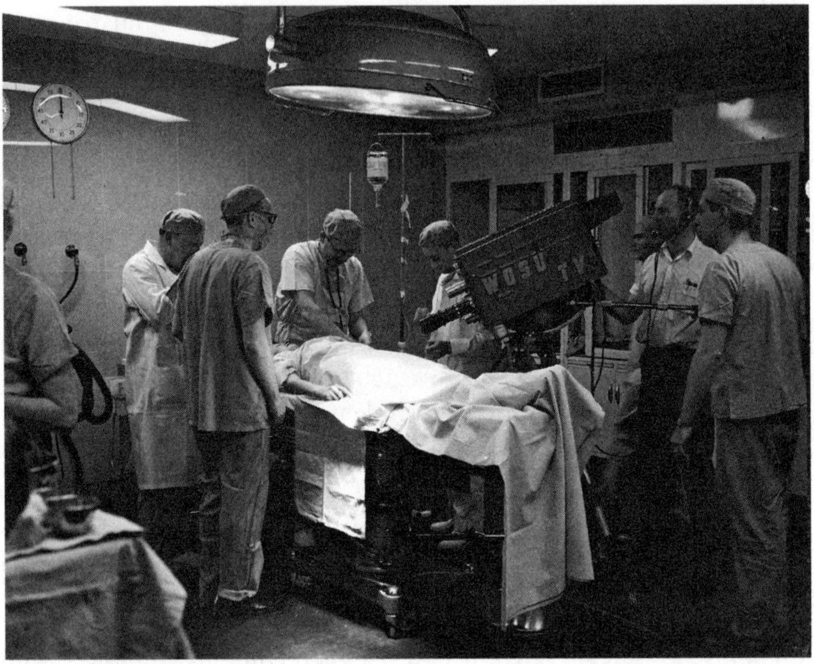

FIG. 12.6. An enormous studio TV camera hovers over a patient as WOSU begins a live video feed to classrooms nearby. Television courses at Ohio State became widely available on campus in the 1960s, with the medical center and the dentistry school providing video programming and commentary for large classes.

for the television courses gradually became less important with the coming of the VHS home videocassette.

GREATER COLUMBUS ETV PROJECT

While WOSU-TV was serving the university with its internal television needs, it also was broadcasting instructional courses to schools as part of the Ford Foundation supported Greater Columbus Area ETV Project. This early attempt used television in second-grade through twelfth-grade classrooms, reaching more than one hundred thousand students. Daytime programming airing on WOSU-TV included courses broadcast to match the class schedules of local schools. The station would fade to black and put up a slide between the course segments, which included French, English, art, chemistry, and health sciences.

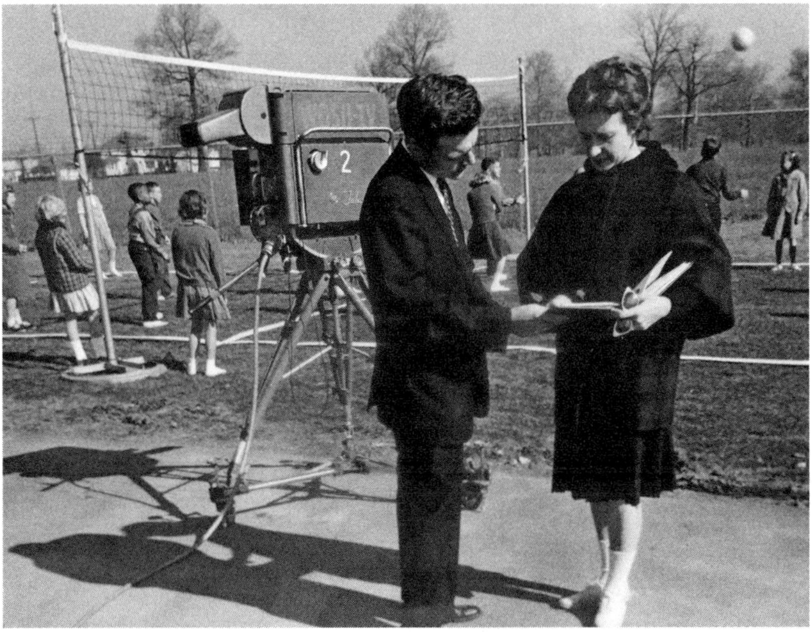

FIG. 12.7. Bonnie Gilliom taught physical education and health sciences for eighth-graders as part of a project with the Columbus schools.

It wasn't unusual for a husband and wife to work in multiple roles at WOSU through the years. Gene Gilliom did occasional color commentary for basketball broadcasts for $25 a game, while his wife, Bonnie, taught on WOSU-TV for several years, with her health sciences and physical education courses sent to schools in the region. The educational programs were sometimes distributed nationally; Gene recalled the surprise of relatives who saw Bonnie teaching on the educational television station in Seattle.

AIRBORNE TV

One experiment in educational television would be considered more than a tad wacky today, but decades ago it earned a $4.5 million grant from the Ford Foundation. In 1961 innovators sent up a converted DC-6 passenger plane to beam programs into schools from 23,000 feet. Serving as a kind of moving transmission tower, the plane would lower a 40-foot antenna to send signals to classrooms on channels 72 and 76. It took off daily from Purdue University and flew a figure eight near Montpelier, Indiana, transmitting some two hundred miles in diameter across six states into the metropolitan areas of Chicago, Detroit, and Columbus. The plane was full to the max, with six and a half tons of equipment, including two television transmitters; two giant reel-to-reel tape recorders; a master control console; and a 75-kilowatt generator in the tail.

In Columbus an airborne educational coordinator was hired, and WOSU-TV churned out short instructional programs using faculty talent that were sent by videotape for airing from the DC-6. Suggestions for teachers appearing on TV, according to one of the consultants, included "[the] mouth and teeth should be normal, without distracting characteristics" and "do not use too many gestures for emphasis; refrain from bobbing or shaking your head." This was well before the days of satellite transmission, and it was exciting to send instructional programs to what was hyped as the "world's largest classroom." When the foundation monies dried up, the flying transmitter in the sky was out of business.

A STATEWIDE NETWORK

As educational television stations attempted to get on the air, funding for the expensive capital needs was partially resolved by federal support. In 1962 the National Educational Television Facilities Act, championed by Senator Bricker, provided matching funds for station construction. It was the first time Congress had given major aid to public broadcasting, and the legislation helped pave the way for the Public Broadcasting Act of 1967.

On the state front, a network of interconnected but independently operated educational TV stations was envisioned with the creation of the Ohio Educational Television Commission in 1961. The agency was charged with ensuring the growth of this "new form of disseminating knowledge called television." Under Dick Hull's leadership as chair, it developed a statewide plan for supporting the expansion of educational television, while advocating for state capital support. In 1969, Ohio distributed $5.5 million to build more television transmitters and equip a network operations center. When WOSU-TV moved from its original facility on North

Star Road to the Fawcett Center in 1971, the state network staff moved in, and the commission has been there ever since.

Dave Fornshell was the first executive director of network operations for the agency, which built a sophisticated microwave system to interconnect all the educational stations in Ohio. In the 1990s, it was replaced by one of the first fiber optic systems in the country to provide statewide program distribution. The agency continued to evolve through the 2000s to become the Broadcast Educational Media Commission, which today distributes state funding allocations to Ohio public television and radio stations and supports them through a centralized TV master control hub.

PUBLIC BROADCAST NETWORKS AND NEW MEDIA (1960–2018)

CHAPTER 13

SPARKS FLEW

> If Mr. Calland was guilty of stepping on the toes of a few stuffed shirts in your organization or was otherwise not a good politician, then he has my backing. I would get rid of the stuffed shirts before taking an action to destroy the university's good image.
>
> —WOSU RADIO LISTENER D. E. KNAPP, COLUMBUS

On July 1, 1960, it was announced that Dick Hull would lead Ohio State's newly established Telecommunications Center. Hull had lobbied four years to create the center as a way of establishing his credibility across campus and incorporating under his purview the radio and television stations, the burgeoning closed-circuit television system, and the Columbus Public Schools television initiative.

The big announcement was somewhat upstaged a few weeks later by a quirky WOSU radio stunt that received campus press. A large photo in the *Lantern* showed staffers Adrienne Hulse and Bob Schweikart at the station's "first annual lawn party." The centerpiece of the tongue-in-cheek event was a charcoal grill set up in a large pothole outside the front door of the radio facility on 19th Avenue. Staffers had become tired of the hole and decided to put it to good use, roasting wieners and marshmallows and playing Christmas music, even though it was summer.

The radio staff event provided a jarring juxtaposition—the erudite Dick Hull growing his domain and seeking credibility across campus from his offices on North Star Road while his ragtag radio staff acted out for all to see on central campus. Don Quayle, who worked in both radio and television at WOSU, described it this way: "The complexity of television, the glamour of the picture, the extraordinary requirement for dollars. It was the new toy that they

FIG. 13.1. In the 1960s, the WOSU radio studios on 19th Avenue still resided in the old WWI aviation building.

didn't know how to handle or how to manage. It was like a tidal wave. I think radio was bastardized by television because they stole all the personnel."

TV was where the action was and where Hull focused his energies. He tapped many of the creative radio staff to host and produce television programs, and some made the switch permanently. All the attention gave the radio employees who were left behind an "inferiority complex" and a "legitimate paranoia," according to Quayle.

Another poke at Hull's domain came in the form of a large, carved wooden sign that appeared in the storefront of the radio offices and announced the ragged place as "Ye Telecommunications Center." Hull's second-in-command, Bill Ewing, sent a memo to all radio staff about the posting of "humorous or 'cute' signs and clippings" in the station. "Mr. Hull and I agree that we must take immediate action," the memo said sternly. The wooden sign was the first to be removed.

SONGS OF PROTEST

It was about this time that A. V. Shirk, who was music director Fred Calland's assistant, put together a broadcast that was not the usual radio fare. The banging and clanging of the station's steam radiators as they came on every cold morning was notorious within the radio operation. "I thought it was an interesting rhythmic pattern, so I recorded it and did a show about an imaginary Caribbean island and locals who had come up with different rhythms, and then I broadcast that radiator noise," Shirk said.

Shirk wasn't at WOSU radio long before he found himself programming and hosting more music shows. While digging into the depths of the music library, he found recordings of black sharecroppers and records by Lead Belly and James "Iron Head" Baker, and he decided to put together a series on the history of black music in the United States. Shirk also started recording music at local clubs with a large Magnacorder reel-to-reel.

Shirk had free rein at the station, because the last thing Fred Calland wanted to do was supervise someone. "He basically let me learn on the job," Shirk said. "The closest thing to management was when he came to me one day and said, 'A. V., are you causing trouble? Maybe I can help!' He was a joy to be around."

Shirk's next project focused on a history of political songs. One of the tunes he recorded in Columbus was a takeoff on a protest song critical of the House Un-American Activities Committee, which had been formed in 1938 and was still investigating citizens and organizations suspected of having communist ties. While putting together the tenth and final program in his series, Shirk included an ad-libbed line from a singer: "Mr. Scherer, are you now or have you ever been . . . a bastard?" Gordon Scherer was a congressman from Cincinnati who served on the committee.

The series aired without complaints until the final show, and then the criticism was immediate and viral. People were definitely listening to WOSU radio. "It was assumed I was a Communist or controlled by the Communist party. I had no idea what I was getting into," said Shirk. Dick Hull called for an investigation. When asked if he had approved the program, Calland backed his assistant and said he had been aware of it. (Actually, Calland had been out of town and knew nothing about the show.) Shirk was suspended without pay for a month, and Calland for a week. According to a memo from Hull, they both had "failed to heed the elementary legal cautions with respect to libel and defamation." Shirk said he was never interviewed about the incident by Hull or anyone else in management. In an effort to make up for Shirk's lost wages, radio staff members held a fund-raising cookout and hootenanny, inviting faculty, staff, and students to "bring your guitars, booze, and Pete Seeger records." They raised $173.50, covering Shirk's monthly salary with $5 to spare.

Shirk continued to help program classical music on WOSU radio, but he never produced another show of his own and left the station in 1964.

CALLAND'S DEPARTURE

In his book about his days in radio, Ken Keller vividly recalled the WOSU facility where he worked in the early 1960s, finding it unchanged in many ways from its 1920 origin: "Never has such a glorious sound emanated from humbler facilities. The deprivations suffered by the staff as a consequence of such ill-housing were many and worth recalling." One problem was the lack of air conditioning. During the summer, windows were thrown open and fans moved the hot air about. Airplanes flying over campus created "such a din through the open windows that announcers or newscasters at work would simply stop reading until they could hear themselves again," said Keller.

The walls of the old World War I hangar were covered with the same material used to make bulletin boards and were full of dents and dings. Several staff recalled the time announcer Gene Gerrard, famous for his quick anger, became so upset with an engineer who played the wrong music to wrap his show that he threw a glass ashtray at a studio wall and it stuck. Doubtless it was removed quickly, since Gerrard and many others were heavy smokers.

Charlie Pickard, who worked at the station as a part-time fill-in announcer for more than thirty years, called the old radio facility "a dump." Pickard found it strange to look up and see all kinds of wires and ceramic insulators from past uses of the space by the electrical engineering department.

In 1964 Hull hired William Steis, a veteran of managing commercial radio stations, as general manager of WOSU radio. Although the quality of their on-air work was not questioned, Hull knew that the radio staff had a party reputation. Hull pushed Steis to make the station "a tighter ship with more professional staff behavior." Keller and others "consistently and deliberately resisted efforts to carry out my directives," said Hull in a letter.

In 1965 Mary Hoffman returned to her hometown of Columbus after living in Europe. Hoffman had received broadcast training overseas and started knocking on the doors of local commercial stations. She recalled the common response: "We don't have any jobs for women on the air; they're all secretarial." When Hoffman landed an interview at WOSU and was told her job would be to assist Calland, she was flabbergasted. She had grown up listening to Calland and thought of him as "a giant" in the community. She started the day after her interview.

By all accounts Fred Calland, classical music director and star announcer, had an unorthodox working style. He would come in at odd hours to program music, and his whereabouts were difficult to track. Sometimes when the station opened in the

FIG. 13.2. A WOSU radio house party in the 1960s. Fred Calland is on the right. Photo courtesy of A. V. Shirk.

morning, Calland was found sleeping on the couch. As Shirk recalled, “Fred had to create a crisis for himself to make things happen. It was a marvel. He smoked, smoked, drank coffee. Stayed up sometimes forty-eight hours at a time, and then he would crash and you wouldn’t see him for days.” There were two camps at the station, according to Hoffman: “One felt that he shouldn’t behave the way he did and he needed to shape up, and others said, ‘My God, he’s our star. We can’t get rid of him.’ ”

Calland’s supervisor, Bill Steis, had decided what camp he was in. In April 1966 Steis fired Calland over a long-term “disagreement over procedures and policy.” There was no debating that for years Calland had been consistently late in preparing the music program listing for the station’s monthly printed guides, which were offered to classical listeners for $5 a year. Calland saw his dismissal as inevitable, saying, “if it hadn’t been the program bulletin, it would’ve been something else.”

Calland's firing led to an outpouring from the community like none other in WOSU history. Hundreds of letters hit the desks of Hull and Ohio State president Novice Fawcett. As one professor put it: "Scratch one more good man lost to the campus—Fred Calland. And who will take his place at WOSU? What is a University radio station without the best?" Among the comments Fawcett received were these:

> From Ava English of Yellow Springs:
> "WOSU has become such an oasis in the desert of commercial radio and TV that I gasp with shock and amazement to think its very spirit may be threatened by the dismissal."

> From Springfield Symphony Orchestra conductor Jackson Wiley:
> "I will be quite frank: WOSU has been one of the things that made me willing to remain and work in Springfield. If this sticks, the single most important contribution OSU has made to central Ohio will be quickly abolished."

Calland raised the level of local ire when he appeared on the WOSU German program he often cohosted and told the audience in German that the station was "going commercial" and would soon be broadcasting popular music. The claim caused a new uproar among classical devotees that played out in the local press. WOSU defended its classical music philosophy but revealed plans to air shorter classical selections interspersed with information during one of the morning hours.

When he left Columbus, Calland cleaned out several thousand classical records from his personal collection that had lived in the WOSU library. He took his talents to a similar position in Amherst, Massachusetts, accompanied by his wife, WOSU radio continuity associate Diana Baker Calland, who had been a talented actress in many of the station's radio dramas in the 1940s and '50s.

When National Public Radio was created in 1970, Fred Calland was hired as its first cultural producer. It was more than a coincidence that the person who brought Calland to Washington, DC, was Don Quayle, NPR's first president. Quayle had left WOSU several years before Calland's firing, and he was so enamored with Calland's talent that he once called him "the one person I knew that I could actually say, 'He was a genius.'" In an interview with Ken Keller, Quayle related an outlandish story about Calland from their days together at WOSU. As the normally unflappable Quayle was anchoring a newscast, Calland came into the studio with the intent of breaking him up and dramatically stripped down until he was buck naked. "I used the 'cough button' more than I ever had before," said Quayle.

Calland had a twenty-year career at NPR, entertaining listeners by producing dozens of live National Symphony Orchestra concerts and other musical programs. Diana Calland became director of radio activities at the Corporation for Public Broadcasting.

A few months after Calland's dismissal from WOSU, Dick Hull sent an optimistic note to his university supervisor reporting that "Dick Howard, Ag Extension radio editor, told me that since Bill Steis had joined the party, WOSU radio looks like a place of business, not a playhouse." Still, the boisterous radio team wouldn't be quelled. They produced an audio blooper reel as a parody about "wonderful WOO-SUE radio." One target was Steis, with one of the announcers providing an impersonation: "This is William Steis. This is B.S. There were many accomplishments over the past year. In fact, I alone issued a record two hundred memos. I got a really big reaction to many of them like this: [dramatic music] 'As of 1967 all part-timers must be available to work on all holidays.' If we had time I would read the other 199."

Two years after Calland left, Bill Ewing wrote to Hull that "whether justified or not, the WOSU image was hurt by the difficulties and publicity surrounding the Calland episode." Ewing said he continued to hear from people who spoke "very critically" about the firing.

Calland was one in a long line of WOSU radio announcers whose knowledge and enthusiasm built a devoted base of classical music fans. His persona would be emulated in some respects by hosts such as John McGrody, Jody Williams, Boyce Lancaster, Christopher Purdy, Jennifer Hambrick, and Mary Hoffman, who replaced him as music programming director. WOSU's announcers also became personally involved with and promoted the concerts of local performing arts organizations, including the Columbus Symphony Orchestra, ProMusica Chamber Orchestra, and Chamber Music Columbus.

A new FCC rule in 1964 limited simulcasts of FM stations by licensees who also owned AM stations. The nonduplication rule forced educational stations with FM stations to develop distinctive local programming, and many began to program classical music. Charlie Pickard saw how classical stations evolved to draw a larger audience over time. "In classical announcing, you used to wear a verbal tuxedo. It was very formal and stuffy, and God help you if you didn't air the entire symphony," Pickard recalled. "Now we play portions of things and quite a variety and more new composers, and the announcers are looser and more conversational."

As one of the few female radio program directors in the country, Mary Hoffman was profiled in 1970. "The radio field is still very much a man's world," said Hoffman. "Men don't necessarily know more about music and programming. It's just something traditional." The writer described Hoffman as "blond haired, gray-green eyed and striking." In her more than two-decade career at WOSU, Hoffman interviewed Russian composer Igor Stravinsky, pianist Van Cliburn, and Baroness Maria von Trapp. She also hosted a daily classical program called *Interlude*.

CHAPTER 14

CAMPUS CHAOS

> April ended at Ohio State with the sad and curious spectacle of those who claim to strive for mankind's peace and dignity and reason, systematically beating other people on the head in the contradictory and self-defeating effort to achieve those ends.
>
> —DICK HULL, MEMO TO WOSU STAFF, APRIL 1970

In the history of educational broadcasting at Ohio State, perhaps only the 1920s could match the frenetic and transformative period that encircled WOSU starting in 1967: WOSU's first color broadcast; campus chaos; a new home; the creation of public broadcast networks; a federal funding source; and more. The period was also remarkable for several major technical accomplishments and a whirlwind of news programming activity. It all would build WOSU into a much stronger radio and television force in central Ohio.

In the fall of 1967 the last major initiative of President Lyndon Johnson's Great Society slate of reforms brought together the hodgepodge of educational television and radio stations into a new public broadcasting network for both. Two years earlier the Carnegie Corporation financed a commission to review the prospects of public broadcasting and make recommendations to the Johnson administration. After extensive testimony, the commission jettisoned the terminology "educational" television for the broader "public" television and set goals to educate, to socialize, to democratize, and to culturally uplift society. The commission created a model for federal funds to support the creation of a network of stations and a national production agency. In a remarkable letter to the commission, still quoted today by public broadcasters, author E. B. White conjured an inspiring image for the future of public broadcasting: "I think television should be the visual counterpart of the literary essay, should arouse our dreams, satisfy our hunger for beauty, take us on journeys, enable us to participate in events, present great drama and music, explore the sea and the sky and the woods and the hills. It should restate and clarify the social dilemma and the political pickle. Once in a while it does, and you get a quick glimpse of its potential."

Educational radio had been left out of the Carnegie Commission's report, but a study funded by the Ford Foundation showed that stations were doing some outstanding programming despite being "underfinanced, understaffed, underequipped, and under promoted." Radio leaders rallied behind the scenes to be written into the Public Broadcasting Act, which was signed on November 7, 1967. The historic day marked the reimagining of public broadcasting as a national force instead of a scattering of individual stations. "Today we rededicate a part of the airwaves—which belong to all the people—and we dedicate them for the enlightenment of all the people," said Johnson.

The act created the private nonprofit Corporation for Public Broadcasting to control the distribution of new federal appropriations. Along with funding that would allow them to develop programs to be aired nationally, stations received community service grants with a formula that in part encouraged local financial support.

The program that would initially define the new Public Broadcasting System (PBS) premiered in November 1969. *Sesame Street* was the result of two years of development and the creation of the Children's Television Workshop, and its strong appeal built a national audience of youngsters and their parents.

Recalling *Sesame Street*'s earliest days on channel 34, Martha Haueisen, the business manager at WOSU, said that when the station couldn't get the PBS feed or had technical issues, calls would flood the WOSU switchboard, many from very young children.

WOSU programmer Merv Durea wasn't impressed by another children's show out of Pittsburgh that he screened in 1969. It featured a genial, middle-aged guy in a sweater and sneakers. "I remember looking at the program and saying, 'I don't understand this at all,'" Durea said. "But then I thought, 'What do I know about kids' programming?' and we picked it up. Of course, *Mister Rogers' Neighborhood* is a legendary success."

CAMPUS UNREST

In January 1968 WOSU radio received a request from Ohio State alumnus Willard Johnson, an air force captain serving in Vietnam. Johnson asked if the station could send him some taped music that wasn't "rock or hillbilly." Staff members compiled fifteen hours of music, including jazz, classical, and folk, and other program material. The effort hearkened back to World War I, when the staff at WOSU radio collected 78 rpm classical records to send to army hospitals.

A few weeks later, it was obvious that the campus had begun to change. Accusations of police brutality were voiced at an Ohio State student assembly after students were arrested during an anti-recruitment demonstration. In the same assembly, black students said they were discriminated against in off-campus housing, course offerings, and student government. The black students soon organized and presented John Bonner, the executive dean of student relations, with eight demands for immediate action by the university. They included the hiring of black faculty and administrators and aggressive enforcement of the university's antidiscrimination policy in off-campus housing.

Responding to the events, WOSU news director Don Davis developed a special series on civil rights issues on and off campus. His interviews with black student leaders Charles Green and David Phears were marked by Davis's spare but direct questioning and the young men's thoughtful answers. It was the calm before the storm.

In early April, as the university considered how to address the student demands, Martin Luther King Jr. was assassinated in Memphis. WOSU covered the rally held on campus the next day. It was a raucous event that displayed a marked change in tenor. One of the speakers said on tape: "The day of nonviolence is over with. It is time for black people to pick up their guns. This bullshit nonviolence is done with King. The only thing The Man understands is violence."

Relations between the black student protesters and the administration deteriorated, with members of what was called the Black

Student Union taking over the Administration Building in late April and holding one vice president in his office for eight hours. After negotiations, the university agreed to a statement that promised progress on a number of the demands. President Novice Fawcett later expelled the eight black students involved in the lock-in, and a grand jury indicted thirty-four students on charges that included kidnapping. The black community and others who supported the students were outraged at the seriousness of the charges, which for the time being seemed to discourage other students from using similar tactics.

WOSU radio regularly reported on the student protests and the responses from the administration. It helped that the studios were only a block from the center of campus and that many of the news assistants were students themselves. The station followed its daily local coverage of the events with a late-night talk program from New York City called *Night Call,* which featured interviews with black leaders such as Ralph Abernathy and Stokely Carmichael which attempted to put race issues in context.

WOSU-TV started a monthly call-in show called *Let's Talk About It* that addressed topics ranging from discrimination to the military draft to the local school levy. Producer Dan Kramer said the station was "striving to get more involved in the community" and to take a stand on social issues. Given experiences with previous audience call-in shows, the station installed a delay system to prevent words that were deemed offensive from hitting the air.

It was a spring and summer of chaos and tragedy in Ohio and across the country. In June riots erupted in the state penitentiary, a few miles south of campus. WOSU's Bill Hurley recalled going up in a prop plane with a cameraman to film the resulting fire at the prison. The inmates rioted again in August and took hostages. Officers blasted through a wall to free them, and five inmates died.

The only balm for the chaotic and controversial year came in the fall when the Buckeyes teed up the pigskin to start the 1968 season.

BUCKEYES IN LIVING COLOR

Buckeye football helped legitimize a tiny educational station in an electrical engineering lab when WEAO radio broadcast the first game in Ohio Stadium in 1922. The situation was strangely similar for WOSU-TV forty-six years later. From its first day on the air, the station had tried to gain viewership by airing Ohio State and high school sports broadcasts. That was nothing compared to the station's coup in 1968, when the Ohio State–Michigan pregame hype was off the charts.

The second-ranked Buckeyes were set to host the fourth-ranked Wolverines on November 25 to decide who would play for the national championship in Pasadena. Incredibly, the national networks already had firm commitments to broadcast other games. The Monday before the matchup, program director Merv Durea gathered the TV staff and announced that ABC Sports needed a local station to broadcast the game for distribution to the Armed Forces Network and to carry it in Columbus. It would be shown exclusively on WOSU channel 34 in living color with equipment provided by ABC.

The station tried everything to promote its first color broadcast. UHF TV converter boxes and antennas flew off the shelves of stores. Columbus resident Mike Shannon, then eleven years old, recalled that his dad gave him an important mission: jump on his bike and hurry to Sun Television and Appliances to buy a UHF antenna to hook up to the family's new color TV. Many other local kids no doubt got the same order. Others may have rushed to Woolco for the store's "Exclusive OSU-Michigan Special," a GE all-channel, 11-inch color TV for $72.

Michigan governor George Romney urged the NCAA to arrange a national broadcast, saying that "thousands of football fans eagerly want to watch the decisive game matching two of the nation's top teams." While a national broadcast wasn't possible, a video feed from Columbus was arranged to accommodate fans in Ann Arbor, where twelve thousand people bought $3 tickets to watch the game on big screens at a campus venue. Fans who lived outside WOSU's coverage area packed into Columbus hotels and local taverns. Both the WOSU television and radio feeds were sent to the Armed Services Network.

ABC brought in the crew and color equipment, and all WOSU had to do was air the broadcast, which was a "clean feed" uninterrupted by ads because of WOSU's noncommercial status.

Head coach Woody Hayes started eleven "super sophomores" in front of a record 85,371 fans at Ohio Stadium. Despite a close first half, the Ohio State offense exploded to win easily, 50–14. After the Buckeyes scored their final touchdown and had a thirty-point lead, Hayes had them line up and try for a two-point conversion. The attempt failed, but it became the most remembered play of the game. When Hayes was asked why he went for two points with such a huge lead, he reportedly said: "Because I couldn't go for three." Buckeye fans appreciated that kind of attitude when it came to their most-hated rival.

Afterward, the steel goalposts came down and exited Ohio Stadium with the thousands of students who marched downtown and, as the *Lantern* reported, "threw a nine-hour party on High Street." Not everyone was in a party mood after the damage was surveyed: fifty windows broken in cars and businesses, and fifty people treated for cuts and bruises.

The Buckeyes went on to beat the University of Southern California and star running back O. J. Simpson in the Rose Bowl and capture their fifth national championship.

It was estimated that more than twenty million people watched or listened to Ohio State–Michigan game via WOSU. According to Dick Hull, that included 80 percent of the TV audience in the Columbus area, or about nine hundred thousand people. The station had been overwhelmed with fifteen hundred letters and more than seven hundred phone calls leading up to the game, most asking how to get the broadcast. And now that many more Columbus homes had UHF sets, antennas, and converters, WOSU-TV became a new viewing option for many residents. It was a memorable way to cap one of the most divisive years in US history.

THE 1970 RIOTS

March 22, 1970, marked the one-hundredth anniversary of Ohio State's founding. The university celebrated with a number of events, including a documentary produced by the photography and cinema department that aired on WOSU. In April the Center for Tomorrow on Olentangy River Road was dedicated on the fiftieth anniversary of the first radio broadcasts at Ohio State. It would soon become the new home of WOSU radio and television.

ACTOR TO NEWSMAN: DON DAVIS

By 1970, Don Davis had moved up the chain at WOSU to become head of TV and radio news. In 1977 he would become radio station manager. Davis credited his career in broadcasting to an injury the summer after he graduated from Columbus Central High School. He was playing touch football in the parking lot of the Big Bear grocery store at Fifth and Grandview avenues when he tripped and broke his wrist. To kill time until he could play sports again, he showed up that summer at Gate 10 of Ohio Stadium to join the university's Stadium Theatre. His first role was as a stable boy because his cloak covered his cast.

FIG. 14.1. Don Davis led a tremendous growth in the WOSU newsroom in the 1970s.

Through the local theatre scene Davis became acquainted with Gene Gerrard

The anniversary celebrations couldn't mask the continuing unrest on campus. Women demanded equal rights, black students sought civil rights and representation, and student protestors demanded an end to the Vietnam War. The board of trustees and much of the university's leadership consisted of older, white, conservative men who couldn't understand and wouldn't put up with student tumult. It was a tense situation, and WOSU was ideally positioned to cover its many angles.

Radio program director Tom Warnock and news director Don Davis had applied for and received the first news production grant from the recently created Corporation for Public Broadcasting. The five-year grant resulted in an incredible growth spurt for the news division, from two full-timers to some twenty full- and part-time producers and reporters. It was now the largest radio news staff of any station in Ohio and allowed WOSU to develop two daily local blocks of news, sports, and weather called *News '70* airing from 11 a.m. to 1 p.m. and 4 p.m. to 6 p.m. *News '70* premiered less than a month before the biggest upheaval ever seen at Ohio State.

On April 29 a student rally attracted more than two thousand people and left the campus literally in tears. The rally was put together by the Ad Hoc Committee on Student Rights, which had presented a series of demands and called for a student strike until the administration negotiated. In the midafternoon, police and protesters clashed at Neil and 11th avenues, where students had staged a sit-in. Columbus police showed up in riot gear and began firing tear gas to disperse the crowd.

At the WOSU studios the chemical wafted through the open windows, causing coughing, sneezing, and watering eyes. Reporters scrambled across campus to gather stories for emergency radio coverage. Davis recalled that one reporter was knocked out after being hit in the head by a tear gas canister. Davis himself was overcome as he tried to cover events on the Oval. "The Columbus police shot off all their tear gas during the day and into the evening, a year's supply in one day," he said. The gas forced WOSU's instructional television offices and the studio in Derby Hall to close up.

WOSU covered the riots heavily, but not without some consternation on the part of administrators. "I know there was a time during the riots that some said we shouldn't be broadcasting at all," said Davis, but Vice President John Mount told university leaders that "as long as they follow the rules and standards and good practices of journalism, I think they should broadcast." It was a rare vote of confidence from the administration.

The confrontations between demonstrators and police continued into the night until Columbus mayor Jack Sensenbrenner declared a curfew for the campus area and the National Guard appeared on

and David Ayers, who were WOSU staff members. Davis and his wife, Tish, later worked as professional actors with a stock company at Lake Erie. After they returned to Columbus in 1956, Davis applied for an announcing job at WOSU and was hired by Ayers. He recalled that he had to cut his hair short before his first meeting with station director Bob Higgy, and that music director Fred Calland "almost had a stroke" when he auditioned but helped him with his voice work.

Davis went on to work at WOSU for thirty years, moving from announcer to reporter to news director to running the entire radio operation. After retiring in 1986, he chaired the Friends of WOSU advisory group.

FIG. 14.2. The 1970 riots on campus shut down Ohio State and WOSU. Tear gas struck a student reporter, and news director Don Davis was overcome by the gas while reporting from campus.

the scene. About fifty people were injured, including eighteen students, twenty-five police officers, and some nonstudents.

Historian Bill Shkurti said the riots were especially complex because they weren't about a single issue: "Fundamentally, this was a revolt against the university. There was discontent from the students during a period when the university had to manage enormous growth. They didn't manage the human side very well. The university worked very hard to keep the lid on in terms of student protests. The problem was, they kept the lid on through the use of force. So when the lid blew, it really blew."

On May 4, on another campus plagued by protests that spring, National Guardsmen at Kent State fired on unarmed student protesters and bystanders. Four were killed and nine wounded. At Ohio State several thousand protesters subsequently targeted a planned Army ROTC awards ceremony. Although insults and obscenities filled the air, the protest ended peacefully. However, on May 6, President Fawcett announced the cancellation of student events due to ongoing skirmishes. He said classes should continue to meet but that the time should be used to discuss the issues dividing the campus.

That afternoon protestors gathered in front of the Administration Building. The National Guard tried to push the students back with rifles and bayonets and were met with bricks and rocks. At 5:30 Fawcett announced that on the recommendation of Governor James Rhodes, he had decided to close the university. The forty-five thousand students on the Columbus campus were told to leave by noon the next day. It was the first time Ohio State had shut down since the 1918 flu epidemic. More than eighty colleges across the country closed that day due to intensifying protests over the Vietnam War and the Kent State killings.

Despite appeals from WOSU leadership, the president ordered the radio and television facilities off the air. "I thought it was a big mistake," Davis said. "We were doing almost continuous coverage, and I thought we were among the most reliable of news sources about what was happening on campus because we were right here. The administration didn't agree." TV broadcast coordinator Don Brown said that the station "more than at any other time provided a valuable communications link between the administration and the university community." Merv Durea said that Ohio State leaders feared protesters might take over the radio or television operations if they stayed on the air—although, he lightly joked, students would never have found the television station since it was so far from central campus.

On May 8 Dick Hull telegraphed the Federal Communications Commission: "This is to advise that WOSU-TV Channel 34 discontinued broadcast service in Columbus, Ohio at 5 PM EDT Thursday, May 7, 1970. This decision . . . was made necessary by an order from the administration of The Ohio State University to close all of The Ohio State University activities. WOSU-TV will remain off the air until further notice. We will advise of any change in our status immediately." It was the first and only time WOSU was ordered to go off the air. Despite the order, some activity had to continue at the television operation because WOUB in Athens depended on a daily daytime feed of educational programming from WOSU through a microwave system.

After the shutdown, Tom Warnock wrote a compelling note to the radio news staff mentioning the many student reporters who had worked long hours and braved the threats of physical violence to cover the situation. "The tone of operations was business as usual in virtually every sense despite the strong waves of tear gas through the building," he wrote. "When news of the shut-down was followed by the directive that radio be included, staff expressed concern and disappointment. Their interest was based on a professional approach and genuine belief in the value of the service they provide."

WOSU returned to the air eight days later on Friday, May 15. *Sesame Street* was broadcast at 7:30 that morning, and the radio operation started airing a mix of news and classical music. The university did not entirely reopen until the following Tuesday. WOSU developed a special program, *Issues and Answers,* to present the many voices of the protests: students, faculty, administrators, and community members and broadcast it simultaneously on TV and radio.

THE OMBUDSMAN

In the fall of 1970 WOSU started an ombudsman project as part of the grant funds received from the Corporation for Public Broadcasting. It was the country's first such service on public radio and would last about five years. Four journalists were hired to research complaints from listeners who had called a newsroom hotline. The team, led by chief ombudsman Don Moffat and executive producer Keith Smith, investigated business deals gone wrong, tenant-landlord disputes, and mail-order rip-offs. They worked aggressively, and Hull had to fend off occasional complaints from local businesses who felt they had been targeted by WOSU. The ombudsman had no legal authority but used what was termed "moral persuasion" along with public support to resolve valid complaints.

The project included as many as six reports a day during WOSU newscasts and drew attention to WOSU as a place where consumers who believed they had been wronged could find a sympathetic ear and perhaps justice. One of the reporters was Anne Gudenkauf, who had started as a student researcher and then worked full-time in the ombudsman program alongside Howard Ornstein, who would later become WOSU news director and then program director.

Gudenkauf also produced the *Access* radio call-in program at WOSU, with topics ranging from gardening tips to the most serious issues facing Columbus. "The newsroom at WOSU was very professional," said Gudenkauf. "I had a great time there and learned so much."

When the Corporation for Public Broadcasting grant ended in 1975, so did the ombudsman project, but Gudenkauf was hired as a freelancer at CPB to write a best practices handbook about what was working at the most-listened-to public stations in the country. In 1977 she joined NPR as the first national editor of *All Things Considered*, the start of a forty-plus-year career at the network.

CHAPTER 15

TIME AND CHANGE

> I am sending this because you are a companion to the lonely and my greatest source of joy and beauty.
>
> —FROM A CONTRIBUTOR IN RESPONSE TO THE FIRST WOSU FRIENDS-A-THON, 1973

Dick Hull had started dreaming about a new broadcast facility almost immediately after joining WOSU in 1956. After brainstorming with a small group of thinkers on campus, Hull sent a report to Ohio State president Novice Fawcett about the potential of a "communications center" that would encompass continuing education, alumni relations, and WOSU television and radio. The report said the center would generate an "in-flow" of thousands of people coming to campus for conferences and an "out-flow" of broadcast information related to those conferences through WOSU.

In 1970 Hull's vision became reality. The consolidated operations of radio and television at WOSU would be housed in the Center for Tomorrow, which was built on Olentangy River Road north of West Lane Avenue, away from central campus and the student population. It included a ten-story guest house with ninety-eight rooms and suites for conference participants and a five-hundred-seat auditorium adjacent to the WOSU facilities.

When Fawcett retired as president of Ohio State, he received a surprise going-away present. The facility was renamed the Fawcett Center for Tomorrow in his honor.

The potential of the facility went unfulfilled from the start. "The center was intended to be much bigger . . . but they didn't have enough money," said Robert Holsinger, a former WOSU staffer and assistant to Fawcett. "So they said, 'Well, we'll use WOSU money to get that wing done, and then we'll use the alumni money to get the other wing done.' Originally the hotel was intended to be much larger, with more meeting and dining facilities. We were millions of dollars short. It would've been a real asset."

WOSU's wing was scaled back, shelving plans for a third TV studio. Still, the radio staff was thrilled to be leaving the World War I airplane hangar after fifty years. As they packed up, someone noticed that the pasty-faced bust of Beethoven in the front window was missing. The window pane had been broken overnight, and there was no trace of the station's unofficial mascot. It was time to move on.

NPR PREMIERES

In 1970 there were four hundred twenty licensed noncommercial radio stations in the United States. More than half of them operated at a power of just 10 watts. Many stations were run by students and couldn't sustain a continuous broadcast schedule. A key challenge of the newly formed Corporation for Public Broadcasting (CPB) was to set criteria for distributing the organization's limited federal funds to stations. The task was assigned to CPB radio director Al Hulsen, who had cut his teeth at WOSU in the 1950s. The initial criteria required stations to have a minimum of three full-time staff, operate at a minimum of 250 watts, and be on the air twelve hours a day. When National Public Radio premiered in 1971, eighty-four stations met the requirements, including WOSU. Others joined as more FM stations came into existence, and today NPR serves as a national syndicator to a network of some nine hundred public radio stations.

WOSU radio had just settled into its new digs at the Center when on May 3, 1971, NPR premiered a daily afternoon newsmagazine called *All Things Considered.* Most people working in public radio agreed with newsman Bill Cohen that the network was a huge blessing for all of public radio. "NPR really invented in-depth journalism on radio," said Cohen. "It ramped up local station quality as well."

TV TRANSITIONS

In July 1971 the North Star Road television studios shut down following the final broadcast from the facility, a live forum with Ohio governor John Gilligan. The moving trucks pulled up to the doors for the short trip to the Fawcett Center for Tomorrow.

The early 1970s was a time of many technological advances for WOSU-TV and radio. After quelling opposition from nearby homeowners, the television station built a 1,000-foot transmitting tower on land the university had purchased north of Westerville. The tower and a new transmitter were paid for through a special appropriation from the state legislature to expand public television access; they nearly doubled WOSU's signal coverage area. In the winter of 1972 new production equipment finally allowed the first regular color telecasts of local programming on WOSU. A year later WOSU assumed direction of a second educational television station, WPBO-TV in Portsmouth in southern Ohio, when the license was transferred from the state educational broadcast agency.

A new WOSU-FM transmitter was installed in 1973 to deliver authentic stereo broadcasts of classical music to the community for the first time. The transmitter also made possible a service offered through an innovative partnership with a new nonprofit called the Central Ohio Radio Reading Ser-

QUBE

For a few years Columbus was the nexus for the future of television. QUBE, which was unveiled on December 1, 1977, had been highly anticipated by everyone interested in the next generation of consumer technology.

QUBE was developed by Warner Communications and introduced a thirty-channel bundle to thirty-five thousand wired homes in Columbus. It was an early form of cable television—and much more. Special handheld QUBE boxes allowed viewers to become involved in the earliest interactive television technology, pay-per-view (including adult movies), and cobbled-together specialty channels for kids, home shopping, and local entertainment. WOSU-TV was listed first on QUBE's remote control at Channel T-2, joined by commercial stations across the region and programming developed at the three-studio QUBE headquarters near downtown.

Although QUBE eventually dissolved because of funding issues in the early 1980s, research on its use in Columbus later spawned cable channels including Nickelodeon and MTV.

vice. The organization started to program a subcarrier channel of WOSU-FM, providing information to special radio receivers designed to pick up the channel. Volunteers read daily news articles and books to benefit the blind and physically challenged in the region. Today, the organization leading the effort is known as VOICEcorps, and the subcarrier arrangement continues.

After the student upheavals on campus in the late 1960s and early 1970s, the WOSU news operation examined its programming and found it lacking in anything produced specifically for the African American audiences of central Ohio. A new show, *Let's Talk About It from a Black Perspective,* started a monthly broadcast in collaboration with Ohio State's Department of Minority Affairs, and a weekly interview show called *Afromation,* hosted by Mary Ann Williams of the Division of Black Studies, went on-air.

A weekend interview program, *Black Studies Journal,* premiered in the 1970s and continued in some form on WOSU radio for more than thirty years. It ended in some controversy in 2007 after a move to a new early-morning time slot caused the longtime host, William Nelson, to resign. Although ratings data had showed a consistent decline in the number of *Black Studies Journal* listeners, Nelson and some friends started a letter-writing campaign targeting WOSU for "racial bigotry." The station defended its record of producing diverse programming with multiple community perspectives, and the issue soon waned. *Black Studies Journal* was the final programming throwback to an era when WOSU simply told an academic department to fill a time slot and ceded any editorial control.

The biggest news of May 1973 was the Senate Watergate hearings in Washington, DC. WOSU took the national feeds and broadcast the morning sessions on radio and the evening sessions on TV until signoff without interruptions or commercials. The station developed a radio documentary, *The Watergate: Facts, Opinions and Ideas,* which included call-in segments about the scandal. Although commercial networks started turning away from the lengthy hearings, NPR and PBS fed WOSU and other stations more than two hundred fifty hours of gavel-to-gavel coverage until the hearings concluded.

Also in 1973, Clay Lowe produced a distinctive program called *High Street* that mixed still images and film to capture the diverse lifestyles of people along the entire length of Columbus's main north-south corridor.

CREATING "FRIENDS"

The 1970s would prove to be one of the most fiscally challenging decades in WOSU history. Every revenue source was under fire, including state, university, and federal appropriations. As the popularity of the WOSU-produced television courses on campus waned, so did the revenue they provided. The university provided 80 percent of WOSU's budget in fiscal year 1972, so when the university cut the station budget by 10 percent, it meant layoffs and serious programming cuts.

The fiscal battering led all of WOSU's AM, FM, and TV stations to reduce their on-air schedules to save on programming, staffing, and utility costs; the television operation turned back the clock and stopped broadcasting any programming at all on weekends, even though studies had shown that most people watched TV on weekends.

The situation was desperate enough that the stations decided to ask their audiences for support. The Friends of WOSU, a fund-raising group led by a community board, was created with the goal of starting a membership club. Merv Durea successfully lobbied for changes in university restrictions to enable WOSU to solicit small-donor memberships on the air. The stations held the first of many Friends-a-Thons in 1973 and raised $24,000 through more than thirteen hundred contributions in a two-day period. Any person who donated at least $15 a year ($5 for senior citizens) officially became a Friend of WOSU. Public stations across the country were beginning to follow suit, and within two years PBS started to help stations by launching its first national pledge drive.

During that initial year of on-air fund-raising, WOSU staff and volunteers were surprised that many fans contributed much more than the requested minimum donation. Notes of encouragement accompanied the cash and checks. One contributor described the "utter frenzy" at the thought of losing some of the WOSU programs, and another called the service a "panorama of electronic goodies." As a direct result of the first on-air effort, WOSU-TV restored programming on Sunday nights, and the FM service brought back classical music programming every weekday morning.

HULL RETIRES

In 1975, after a nineteen-year reign over WOSU, Dick Hull retired at the age of sixty-one. He was a heavy smoker, and the habit had caused serious health issues. Despite his issues with the radio staff, Hull had built WOSU into a strong, respected organization, and few administrators ever questioned his authority. During Hull's tenure, there were no internal reviews, and no university committees monitored WOSU.

No other leader in WOSU history had attained Hull's national stature. As radio and television producer David Ayers said, Hull always thought big: "Some may not have fully understood his vision, but they knew he had one." Dave Fornshell, who ran the state educational broadcasting agency, pointed to the valuable work of Hull and Keith Tyler in convincing the FCC to reserve channels in each state for educational television. "Without this action, very few, if any, educational TV stations would exist today," wrote Fornshell. Throughout his career, Hull made sure the light of educational broadcasting continued to burn brightly.

NEW LEADERSHIP

To replace Dick Hull, Ohio State hired thirty-two-year-old Elizabeth Young in 1976. Young had worked in radio programming before moving to Washington, DC, where she was mentored by two former WOSU staffers. Al Hulsen at the Corporation for Public Broadcasting hired her to assist with station activities; then Don Quayle brought her to NPR to run station relations. Young was also chief consultant for a study on the role of women in public broadcasting.

With a new director at the helm, Ohio State decided to review the effectiveness of WOSU and its relationship to the academic mission of the university. The thirteen-month study by a faculty and staff committee resulted in a generally positive report, although it suggested putting a lid on Ohio State's investment and encouraging

FIG. 15.1. Elizabeth Young, the third director in WOSU history, cohosting the Festival '77 fund-raiser with WOSU's John Holscher.

the stations to seek much more private support. The report credited Young's early efforts to involve more students in the organization. "We feel we are fulfilling our obligation to the students," Young said. But she also emphasized that WOSU "was not conceived of as an academic training ground, but as a service to the community."

Young took over the one-hundred-member staff at the Fawcett Center just as the country was coming out of a two-year recession. She focused on expanding the membership base by working closely with the new Friends of WOSU Board and bringing more attention to the stations. If WOSU could find a way to raise substantial community support, it could flourish beyond the absolute control of its licensee. What WOSU leaders didn't know was whether audiences would continue to contribute to support stations they could watch or listen to for free.

That question was answered most enthusiastically during Festival '76, the first extensive effort at WOSU to use special pledge programming to raise funds. It brought in $36,000 by airing PBS offerings *Richard III* and Ingmar Bergman's *Scenes from a Marriage.*

That same year, despite its own fiscal challenges, WOSU attempted to help another prominent nonprofit with serious funding issues. It requested a waiver from the FCC to allow its FM and TV stations to broadcast a fund-raising auction, with all proceeds supporting the Columbus Symphony Orchestra. The FCC denied the request, saying the broadcast would be inconsistent with the noncommercial nature of educational broadcasting.

In 1978 the Friends Board created an endowment fund with the Ohio State University Foundation with more than $100,000 from a number of local corporations. By 2018 the fair value of the endowment contributions and investments had grown to more than $5.4 million.

The fundraising efforts continued and Festival '78 WOSU raised more than $100,000 from listeners and viewers. To celebrate the first twenty-five years of educational television, live entertainment with celebrities including Hugh Downs and Julia Child was fed from WNET in New York City. Volunteers from the Friends Board manned the telephones, logging more than five thousand pledges. A special classical music event that year packed Mershon Auditorium as the Ohio State Symphony Orchestra

CRACKING THE GLASS CEILING

With a female director, Elizabeth Young, WOSU was a rarity among the nation's broadcast stations in 1976. The station also had hired one of the few female engineers in the business. Tessa Unwin was the only woman among ten engineers at WOSU radio; she handled a variety of audio production duties, including technical support for *Access,* a live call-in program. In an article after her hiring, Unwin said she felt accepted and welcomed by her engineering colleagues at the station.

Thirty-three years later it was still unusual for broadcast stations to have female engineers. That was reflected in the publicity WOSU received locally and in the national broadcast press after promoting Ling Ling Sun to chief engineer for television in 2009. Sun had grown up in China in the 1970s and had read about television but had never seen an actual TV set until she was eighteen. At the Beijing Broadcast Institute, she found joy in "seeing a problem, fixing it and seeing the result right away." Once she made it to the United States, Sun worked at the public station at Washington State University, and she later found a home at WOSU.

Sun was a vital part of the small team in 2006 that helped design and advise on the renovation of the WOSU studios at COSI, the science and industry educational center in downtown Columbus. Sun was at WOSU for nine years before being promoted to the top engineering spot. After three years in the position, she was hired by the Nebraska Educational Telecommunications as chief technology officer.

presented its opening concert with guest conductor Karl Haas, the popular host of the national program *Adventures in Good Music.* The event was broadcast on both WOSU-FM and WOSU-TV as part of the annual September Festival to raise membership funds.

The number of days of on-air pledge programming continued to increase, even as station leaders admitted that the efforts detracted from their primary mission of providing outstanding programming. In the 1970s the main goal at WOSU was survival.

In the fall of 1978 Liz Young and a group of guests dedicated a new 10-meter satellite receiving dish behind the Fawcett Center. The dish, which replaced an expensive AT&T landline, enabled WOSU to receive up to four PBS programs simultaneously. Young left WOSU a few months later to become president of the Public Service Satellite Consortium, which coordinated the satellite system for PBS. Young said that in her three years at WOSU—an unusually short stint for the station's leaders—she was proudest to have increased memberships fivefold and brought more women and minorities onto the staff.

The year 1978 would be remembered for a late-January blizzard, the worst in Ohio history. The storm raged for over two days with gusts of nearly 70 mph and wind chills of 40 below zero, bringing central Ohio to a standstill. Howard Ornstein recalled News Director David Feingold calling him at 4 a.m. to come into the station. When he finally made it, all the power was off and he had to feel his way through the building to the basement newsroom. The radio operation used a generator and a portable audio board to broadcast the news. "People just started coming in and we had our whole news staff here," said Ornstein. "It was like camping. We slept in the Fawcett hotel rooms and would roll out of bed and go to work downstairs. It was serious work during the day (51 people died in Ohio due to the blizzard), but we had a toga party one night and had a good time. It really brought the staff together." Because there was no electricity, the marooned staffers were allowed to help themselves to food in the Fawcett Center restaurant before things could spoil. Merv Durea recalled reading the news by candlelight and accidentally setting his hair on fire. The blizzard stranded reporter Bill Cohen for several days at the Statehouse, where he covered Governor James Rhodes's numerous press conferences and slept on a bench outside Rhodes's office.

FIG. 15.2 *Bluegrass Ramble* celebrates 40-years on WOSU radio with a special concert in September 2018 that included *Slate Ridge* with Ramble originator Tom Ewing. From left to right: John Obora, Rich Baker, Dan Cook, Tom Ewing, and Steve Brechter.

BLUEGRASS RAMBLE

As early as the 1940s, WOSU brought in musical acts unique to Columbus radio from the foothills of Appalachia. The fiddlers and banjo players sparked interest in the music that would become known as bluegrass. A student performer, Dana Vibberts, had a daily show in 1951 called *Folk Trails* on which he aired early music recordings, including bluegrass, and he himself performed. Vibberts later joined the faculty of Ohio State's photography and cinema department and performed periodically on WOSU radio into the 1970s.

In 1963 the Sacred Mushroom on High Street featured the Country Gentlemen, promoting the music as "folk music with overdrive," a common shorthand description of bluegrass. The 'Shroom, as it was called, was one of several bars and clubs that hosted bluegrass and folk musicians. Many of the clubs could be found on the south side of Columbus along Parsons Avenue. They attracted blue-collar workers from the nearby steel mills and factories as well as college students from around town.

By the 1970s bluegrass groups regularly included the campus area on their tours. Doc Watson performed at Hitchcock Hall Auditorium; Earl Scruggs brought his banjo pickin' to the Agora; and John Hartford played at Zachariah's Red Eye Saloon. WOSU-TV promoted a local studio special called *Sing of the Times,* featuring bluegrass, mountain, protest, and folk music.

In 1978 Tom Ewing, a local teacher and bluegrass musician, heard that the station wanted to expand its music programming beyond classical. "I had a bluegrass record collection and I thought I'd let WOSU use it. The next thing I knew I was being auditioned to host a program that debuted on July 6th, 1978." The new show was called *Bluegrass Ramble.* After Ewing joined Bill Monroe as a member of the Blue Grass Boys in 1986, Rich Baker took over as host for "the Ramble." Baker has been the most consistent host with nearly thirty years at the microphone, but many others have held the gig, including Chet DeLong, Steve Brechter, and Chris Johnston; WOSU staff member Brent Davis was a regular substitute.

After receiving a license to broadcast after dark in 1988, the AM station aired *Bluegrass Ramble* from 6 p.m. to midnight on Saturdays and Sundays. Knowledgeable hosts filled the time slot by taking requests, listing upcoming concerts, and telling stories about the artists. *Bluegrass Ramble* has always served a diverse audience, from faculty members to factory workers. Brent Davis recalled a station tour during which a Somali immigrant to Columbus said he listened to the show because the themes of loss, love, and yearning for home in the music reminded him of the songs of his culture.

In 2011, when WOSU sold its legacy AM 820 station, *Bluegrass Ramble* moved to WOSU-FM 89.7 as the only music programming on an otherwise all-news station. Its consistent success during radio pledge drives and steady audience numbers have kept bluegrass music alive on radio in Columbus for more than forty years.

CHAPTER 16

GROWING PAINS

> The university funding cuts won't kill WOSU, but it will change WOSU. Fund-raisers and private support are even more critical now.
>
> —DALE OUZTS, 1982

In 1975, before the hiring of Elizabeth Young, Ohio State brought in a team of public broadcast consultants to help identify the challenges ahead for the next head of the Telecommunications Center. One of those consultants was Dale Ouzts, a broadcast veteran working at what was then Sangamon State University in Springfield, Illinois. Four years later, Ouzts was back in Columbus after a stint with NPR, following in the footsteps of Young, Dick Hull, and Bob Higgy as the fourth leader—and the first with the title of general manager—in WOSU history. With the hiring of Ouzts, the broad electronics communication purview of the Telecommunications Center began to fade. The focus shifted solely to the radio and television channels, which would become known as the "WOSU Stations."

Ouzts, who grew up in Geneva, Alabama, promised in a pronounced southern drawl to make WOSU "lean, mean, involved, and aggressive." The statement was typical of the brash Ouzts, who was known for his droll sense of humor. When asked about his accent, he would often say he was from "L.A." and pause for effect before adding, "That is, Lower Alabama."

Over the next two decades Ouzts would lead WOSU to a number of accomplishments, but also would be buffeted by controver-

FIG. 16.1. Dale Ouzts (center) with long-time WOSU staffers Phyllis Madry and John Prosek.

sies, missteps, and fiscal challenges that dominated his tenure. While developing his first budget, Ouzts learned that the station's financial situation was in dire straits. In June 1980 he gathered his staff for the bad news: "We had a meeting in the Fawcett Auditorium, and I read the names of twenty people we had to lay off," he recalled. "We didn't want to do it, but we had to."

Despite the difficult start, the radio operation began to successfully transform itself in the early 1980s with new identities. The AM station became News 820 and the FM station became Classical 89.7, establishing separate formats of news/talk and classical music for the first time. Research showed that stations with a variety of formats were becoming a remnant of the past; listeners wanted to know what type of programming they would be getting when they tuned in. The format changes and new promotional strategies paid dividends over the short term with strong audience growth.

The News 820 lineup was driven by two NPR signature programs and a strong local news presence. *Morning Edition* had premiered in 1979 with Bob Edwards as host, joining the estab-

FIG. 16.2. The WOSU news team with station management in front of the Fawcett Center for Tomorrow in 1982.

lished afternoon magazine *All Things Considered.* Carl Kassell gave the national news on the show, and popular guest commentators included Red Barber, Frank Deford, and philosopher and large-animal veterinarian Baxter Black. News 820 cycled *Morning Edition* from 7 a.m. to noon and then filled a four-hour gap with local programming until *All Things Considered* hit the air. In its first year, WOSU-AM saw average weekly listenership grow by an astounding 50 percent to eighty-four thousand.

WOSU also unveiled a new program guide in 1980 to replace *The Broadcaster,* its monthly tabloid newspaper. *Airfare* was designed to be an attractive, slick magazine that would look good on a living room table. It has changed size and scope many times through the years, but four decades later it remains WOSU's monthly program guide.

WOSU partnered with Ohio Educational Broadcasting, the agency that connected all of the state's public stations, to build a network for radio news coverage of state government. With five years of experience covering the Statehouse, WOSU's Bill Cohen helped set up the bureau and continued to report on that beat for another thirty-three years.

Ouzts began to build a relationship with the Beijing Broadcasting Institute, and WOSU often hosted Chinese teachers who spent four months on campus learning modern broadcast techniques. Chief engineer John Battison assisted with the project, completing a distinguished career working at WOSU. Battison had come to the United States after serving as a World War II bomber pilot with the Royal Air Force. His professional achievements included more than one thousand published articles on technical broadcast issues.

Battison and Ouzts first traveled to China in 1981 to give a series of lectures at the Beijing Broadcast Institute. During one of their trips, Ouzts was joined by an eight-member production team from WOSU. With an assist from a federal grant, the team spent nine weeks capturing events, including performances of an acrobatic company and the national orchestra. TV producer Fred Andrle recalled how some of the team returned to Columbus tired and gaunt from the twelve-hour workdays and their distaste for the local food. WOSU turned the mounds of videotape into six nationally distributed documentaries.

HIGGY'S CODA

Bob Higgy, whose influence had shaped WOSU from the beginning, passed away in 1981. Higgy's career started with his fascination with the telegraph. His deft touch on the key and his ability to extend the power of his homemade transmitter led to his being considered one of the stars of amateur radio. It was no wonder the faculty of the electrical engineering department turned to the "genius radio operator" and hired him after graduation. Within a few years, the twenty-six-year-old was running the station. For the next four decades, Bob Higgy was synonymous with WOSU radio. As his daughter Joyce said, "Radio was his mistress."

In the 1960s, Higgy was diagnosed with Parkinson's disease. Despite experimental treatments at the Ohio State Medical Center, the disease progressed. Higgy's mind was active but he was eventually housebound, unable to verbalize clearly. His daughter described how his only visitors were longtime ham radio friends. They would set up a telegraph key and Higgy would slowly tap out his thoughts, letter by letter, word by word, in the dashes and dots of Morse code, a form of communication he had mastered by the time he was ten.

Higgy died of complications of Parkinson's at the age of seventy-nine.

FIG. 16.3. Brian Naylor and Lynn Neary (shown here) along with Anne Gudenkauf, Tamara Keith, Eric Nuzum, and others cut their teeth on WOSU radio before starting long careers at NPR.

RADIO TALK

Lynn Neary worked at WOSU for less than three years, but she marveled at the commitment to local news. She had the education beat but also locally hosted *Morning Edition* and *All Things Considered* during her tenure. For a time she even cohosted a weekly television show with colleague Brian Naylor called *Columbus Perspectives.* Neary won a statewide radio documentary award for a series on the problems facing Vietnam veterans and recalls profiling Marysville, Ohio, for NPR before the giant Honda auto plant was built there.

WOSU had experimented with regular radio talk programming since a 1974 partnership with the League of Women Voters led to the creation of a public affairs show called *Metro Dial.* The *Access* call-in series started in 1977, airing several times weekly, but in March 1982 the station took a major plunge into talk radio as a daily talk program was christened. The program would have multiple hosts and iterations through the years, but its first was Lynn Neary, who gave the program its name, *Open Line.* She would have little time to dig in and develop *Open Line* as NPR came calling, and she left within a few months to become a reporter and anchor for *Morning Edition.* She was joined in Washington a few months later by Naylor, who helped cover the Ohio Statehouse for WOSU.

It's notable, given WOSU's history of showcasing female announcers, that there was effectively a ban on women's voices by commercial network radio from the 1920s through the 1970s. NPR disrupted the tradition of the baritone broadcaster by featuring Susan Stamberg's signature sound and the voices of Linda Wertheimer, Nina Totenberg, and Cokie Roberts, who were joined through the years by former WOSU news staffers Neary, Gudenkauf, and Tamara Keith, who was a reporter at WOSU from 2004 to 2006.

THE DEATH OF A PRINCESS

John Prosek had just been hired when WOSU shook up the staff with layoffs in 1980. The station then hemorrhaged talented producers after a management reorganization, giving Prosek the opportunity to move from master control to director to producer to production manager. As a director, he developed *Columbus Perspective,* later renamed *Columbus Week in Review.*

Prosek related an incident early in his career from 1980 when PBS planned to show the controversial docudrama *The Death of a Princess,* which was based on the true story of a married Saudi princess and her alleged lover from a common class. The romance was revealed, and the two were executed for adultery, as dictated under Islamic law. The planned airing roused international furor. The Saudi government pressured the US State Department to intervene with PBS, and Mobil Oil placed ads objecting to the broadcast. While damning the film, the Saudi royal family never denied that the execution took place.

Thirty-eight PBS stations in thirteen states chose not to show the program. Most PBS stations, including WOSU, aired it as scheduled at 9 p.m. on May 12, despite a number of callers who seemed most concerned about the impact it might have on oil imports from Saudi Arabia. A handful of protesters gathered outside the Fawcett Center, and several police officers were on guard inside. Prosek recalled a crew from WBNS-TV rolling tape for a news story in the control room as he put the docudrama on the air. "My only concern was that after it aired, everyone up and left and I was alone getting out of the building," he said. Prosek was left unscathed, as was WOSU's reputation.

1980s WOSU TV

By far the longest-running local show on WOSU-TV and in Columbus is the high school quiz tournament called *In The Know.* After years on WBNS-TV, the show moved in 1983 to WOSU, where it was hosted by a smart, witty certified public accountant named Bill Schiffman. Also in the 1980s, a local TV special, *Blacks and the Constitution,* premiered with moderator Norma Quarles and won national awards for excellence. WOSU partnered to produce and distribute a profile of Warren G. Harding called *The Front Porch President.* In 1989 WOSU broadcast the dedication of the $43 million Wexner Center for the Arts on campus. The star-studded event featured Tony Award–winning actress Colleen Dewhurst, composer Philip Glass, and journalist Barbara Walters.

WOSU-TV aired old reruns of *The Twilight Zone* with Rod Serling, who had Ohio connections, having attended Antioch College in Yellow Springs and worked as a writer at WLW in Cincinnati. The show drew a good audience at 11 p.m. weeknights against local commercial news. The station also programmed 1960s reruns of

FIG. 16.4. WOSU's Dave Nicholson on camera at Ohio Stadium. The OSU maintenance truck hauling the camera along the sideline almost tipped over. It would not be used again.

Alfred Hitchcock Presents. By 1982, due to fiscal issues, the station dropped both of those syndicated programs and chose to stick to less costly and traditional PBS programming. Other cutbacks that year included signing off the TV station at 11 p.m. and reducing transmitter power to save on utility costs.

Perhaps the most popular local programming on WOSU-TV was the airing of Ohio State football games. From 1978 to 1990, WOSU was allowed to produce every Buckeye game for delayed broadcast. The games—most of which were played on Saturday afternoons—were taped and then aired at 10 p.m. on Saturdays and again on Sunday mornings. Each broadcast included a crowd favorite, the Ohio State Marching Band's pregame and halftime performances.

Commercial network broadcasts of Ohio State football were still rare in the 1980s, so the delayed games and band broadcasts became a popular feature. The productions were elaborate for the time, with five cameras, multiple replay machines, and a crew of twenty-seven. For most of those years Ed Clay was executive producer and Prosek was director and producer, with announcers Jack Kramer and Cleveland Browns football legend Paul Warfield.

BIDDING FOR BUCKS

Although WOSU had successfully grown its membership base with the pledge programming initiated in the 1970s, it decided in 1984 to follow the lead of a few other PBS stations and hold an on-air TV auction to raise more operating funds. For the next several decades, the station devoted fifty hours of live prime broadcast time to the auction during the last week of April and first week of May. The event annually grossed $150,000 to $300,000, with local hosts auctioning off everything from autographed photos of aged television stars to a Honda motorcycle. One year the station auctioned a DeLorean car for $30,000. Hosts included Bill Schiffman, who served as auction chair, and a number of WOSU and local radio personalities.

Columbus Zoo director Jack Hanna was a regular guest during the initial auctions and brought in a variety of animals to entertain viewers, a model he famously followed with his appearances on the *Tonight Show* with Johnny Carson. Some of the Hanna stories are legendary around the station. Guest host Ken Keller recalled having a parrot poop down the back of his tuxedo while he was on camera. Many of the Hanna visits involved a snake, including this story from Fred Andrle:

"Hanna came out and put a big python under one flap of my sport coat. I could feel it coil around, and its head came to rest just

FIG. 16.5. Jack Hanna of the Columbus Zoo was a regular guest of the WOSU auction. Note the snake around the neck of general manager Dale Ouzts.

below my neck and the rest of the body in my arms. A minute later he brought a bear cub on and it took a swipe at the snake, and Jack pulled the bear back just in time. Later I asked Jack what would've happened if the cub would've hit the snake, and he said, 'You would've definitely been bitten by the python.' Those were the days."

While the auction was moderately successful, it involved an immense amount of staff time, including the coordination of hundreds of volunteers, and net profit was minimal. After twenty-five years the WOSU-TV auction was taken off the air in 2008.

DAMAGE CONTROL

The Ohio State president's office still supervised WOSU in 1987, but that changed after university administrators decided to take a much closer look at the station's operations. The decision may have been prompted by the television auction, which had drawn negative attention in the press the previous year. Headlines pointed to the use of shills or fake bids to jack up value and increase bidding on items. The story gained prominence in the *Columbus Dispatch,* although the volunteer who made the accusation later said he got the story wrong. In the end WOSU representatives admitted to making some bogus bids, but only for a few dollars above the minimum on slow-moving auction items, and none of the items sold.

A short time later President Edward Jennings ordered a formal internal and external review of WOSU. It was only the second nonacademic unit review in the university's history. The process stretched beyond two years, and leaks to the press spawned a series of negative headlines in local papers and magazines, including Ohio State's student newspaper, the *Lantern* ("WOSU breaks own code in production of show"), and in Cleveland's *Plain Dealer* ("OSU TV, radio stations ripped in reports").

The internal review team, which included faculty and a few representatives from WOSU, criticized station leadership and cited poor staff morale and low salaries. The study also focused on lapses in ethics, including the production of a television series paid for by Mount Carmel Health, a local health-care system, that was the type of "pay to play" programming deemed unethical by PBS standards.

External reviewers, including public broadcast leaders from Seattle, St. Paul, and the University of Illinois, came to campus following the completion of the internal review. Their report blasted the internal process as "seriously flawed," saying that it should have included a self-analysis developed by WOSU; that station staff members had insufficient involvement in committee deliberations; and that the lengthy period of scrutiny and debate led to a report lacking any unity of opinion.

The visiting broadcasters recognized WOSU's long history of accomplishment and praised many aspects of the radio news and classical music operations, but they were pointedly critical of the TV operation, calling the station output "basic at best" and encouraging a stronger commitment to local TV programming. One of the evaluators said it was clear WOSU was "a radio operation first."

They also were concerned about attempts to control or influence the editorial decision making at the station. In one of several cases involving administration meddling, a special assistant to President Jennings prodded Ouzts to air the memorial service for football coach Woody Hayes in prime time in place of a planned fund-raiser. Ouzts told the *Plain Dealer* he had little choice but to comply: "I was told I wouldn't be here if I didn't [air the program]. I had two kids in college. What was I supposed to do?"

The interference in station programming perturbed the reviewers, who wrote that if WOSU were to achieve excellence, "an essential element would be the public's perception that the station's programming has editorial integrity."

Publicity surrounding the reviews embarrassed Jennings, and he decided to wash his hands of WOSU, shifting oversight to the dean of the College of Social and Behavioral Sciences, a division that included communications and journalism.

The review process that started in the winter of 1987 finally concluded three years later. The external reviewers commented on the enormous stress the lengthy process had put on the WOSU staff: "Not one member . . . was unaware that the organization was under the microscope, being scrutinized and that a mistake in judgment could lead to an unfavorable report which could damage the station." There was substantial staff turnover over the course of the review, including the loss of eight managers.

As a response to the review, WOSU leadership created a *Plan of Action for WOSU Stations*. The strategic plan outlined steps to tackle most of the review's recommendations and was endorsed by Jennings, Ouzts, the provost, and the dean of social and behavioral sciences. It established a new mission statement, which held that WOSU "will always seek to attain the highest levels of ethics and editorial integrity." It also called for the station to develop a long-range strategic plan, begin formal leadership evaluations, improve internal communications, and give staff a greater role in setting priorities and goals. Funding would be sought to help increase salaries and training opportunities, and the university agreed to work with WOSU to resolve space issues at the Fawcett Center.

The many endorsements attached to the plan seemed to indicate that it would hold some weight, with deliverables tracked in future years. However, despite its heft, the plan seemed to have limited impact. Jennings left the university within months of the report, and the college dean who supervised WOSU retired two years later.

The reporting line to a dean of an academic college ultimately brought only more frustration. WOSU's annual reports over the next decade consistently showed the chief challenges of the organization to be the lack of funding, staff, and space. And in most cases the blame was put at the feet of college dean and university administrators.

When budget reductions hit the university in the early 1990s, WOSU took the brunt of the cuts within the College of Social and Behavioral Sciences, and the college's strategic planning document barely mentioned the station. There continued to be little clarity about WOSU's role at Ohio State, and the gulf between the station and the university widened.

TONGUES UNTIED

In 1989, the station was back in the news, this time in the company of public stations across the country as they wrestled with an award-winning documentary depicting the lives of gay black men called *Tongues Untied.* The program was part of a PBS summer series under its *POV* (Point-of-View) brand. A number of stations—including those in Denver, Houston, Milwaukee, and South Carolina—refused to air *Tongues Untied* because of images of nudity, the use of profanity, and the theme of homosexuality.

Ouzts and WOSU's Ed Clay, one of the few black television managers in the country, reviewed *Tongues Untied* and decided it presented a provocative point of view the community should be allowed to see. They also decided it should air at 1 a.m. and six weeks after its scheduled premiere on PBS.

Ouzts saw the issue as a no-win situation for managers: "If you show it, you're a pornographer; if you don't, you're a censor." WOSU received at least one hundred letters and a number of phone calls about the decision. Reactions ranged from shock that the station would air "such a vile and disgusting" program on the same network that aired *Sesame Street* to indignation on the part of people like Dennis Toth, film coordinator at the Columbus Museum of Art, who said the 1 a.m. airtime was "supportive of repression and censorship and not in the spirit of public broadcasting."

A year later WOSU was again accused of censorship when administrators refused to allow *Gaybeat* magazine to fund its radio programming. *Gaybeat* wanted to buy time on 820 AM with this statement: "News 820 is supported by *Gaybeat,* Ohio's serious and fun gay newspaper now headquartered in Columbus." Ouzts claimed accepting the support would be similar to allowing political parties and antiabortion groups to underwrite programming. The publisher responded that *Gaybeat* was a magazine and not a social issue. Editorials in local newspapers called the action discriminatory, and Stonewall Union, the gay and lesbian support organization, said the WOSU policy was homophobic. "Sexual orientation is not a controversial issue," wrote Stonewall leaders. "One's orientation is innate and not a choice." Ouzts said the station was not denying *Gaybeat* anything: "We just aren't taking [its] money." Under FCC regulations WOSU could refuse funding from any organization for program support. The station stuck to its policy, and Ohio State leaders, including new president E. Gordon Gee, deferred to Ouzts.

That same year, WOSU brought suit against American Public Radio (APR) and competing local public station WCBE-FM over carriage rights to *Garrison Keillor's American Radio Company of the Air,* the successor program to *Prairie Home Companion.* WOSU claimed the show was a new offering and, as the primary APR affiliate in the market, it was entitled to broadcast the program. WCBE claimed the rights because it had been carrying Keillor's *Prairie Home* program exclusively in the market and the new program series was simply a continuation of that program. A judge awarded the program to WCBE.

ELIMINATE WOSU?

When Ohio governor George Voinovich threatened deep cuts in the state's contribution to higher education in 1992, WOSU was among the areas that made Gee's quick hit list. Gee said WOSU, as well as

The Ohio State University Press, would be shut down if the reductions were enacted, and the university would raise tuition by 25 percent.

The uproar that followed was tamped quickly by the university's public relations office. Officials announced that there were no plans to get rid of WOSU or cut off its funding, which amounted to $1.3 million or about 20 percent of the overall station budget. Ultimately, state cuts to higher education would be minor.

Gee's offhand remark captured the attention of Tim Feran, who covered media for the *Columbus Dispatch*. Feran wrote a column suggesting that if Ohio State kept cutting support for WOSU, the university should simply turn the station over to the community. He noted that Seattle's public television station had moved from being a university licensee to becoming a successful community-led station.

The Feran column and internal discussions at WOSU led to an independent study by Paul Krause and Associates, a Minnesota-based firm that specialized in economic analysis of station-licensee relations. The study—which ultimately helped WOSU retain its university subsidy—estimated the benefit of the station's airtime devoted to Ohio State at nearly $2 million annually. Simply the use of the WOSU call letters powerfully extends the influence of the institution well beyond campus. The study didn't include many of the other production and support services provided by WOSU or the national exposure the university received through programming such as *Body Talk,* a popular one-hour call-in show that aired on dozens of public radio stations across the country. Hosted by Tom Wiebell and produced by Eric Nuzum, *Body Talk* featured a rotating panel of Ohio State physicians, including the popular Robert Guthrie, who provided detailed answers to call-in questions.

MIXED BAG

In the early 1990s PBS viewership across the country reached new heights, assisted by *The Civil War,* the documentary series produced by Ken Burns. About forty million viewers saw the series over five nights, making it at the time the most-watched program in PBS history.

Despite the national success, funding issues at stations like WOSU meant a ramp up of fund-raising efforts to cover continued decreases in state and university support. A record ten fund-raisers were held between the three stations (AM, FM, TV) in 1993, along with the annual television auction. WOSU started to invest in major and planned giving programs and also announced a new endowment campaign. Some positive news came with a $500,000 bequest from the estate of Don Glancy, a professor emeritus of theatre at Ohio State and longtime fan of Classical 89.7. At the time, it was the largest donation from an individual in WOSU history.

Meanwhile, the college dean assigned to supervise WOSU, Randall Ripley, prodded the station to create more programming that highlighted the university. In the following years WOSU certainly delivered. An Emmy Award–winning documentary, *CsurisVision* (1995), profiled Professor Chuck Csuri, widely acknowledged as the father of computer animation. To celebrate the university's 125th anniversary, the station aired vignettes highlighting campus history and produced under contract the Ohio State historical documentary *Echoes across the Oval* (1996). *The Birth of Ohio Stadium* (1999) focused on the innovative architecture and construction of the massive Horseshoe. WOSU continued the long tradition that started with radio—the tradition of capturing for broadcast all Ohio State commencement exercises and every State of the University speech.

CHAPTER 17

BEYOND OSU

I always wanted to do talk radio. It was on my agenda even as a kid. I have direct access to listeners and never know which direction the conversation will go.

—FRED ANDRLE, LONGTIME HOST OF *OPEN LINE*

The popular *Open Line* radio talk program, first hosted by Lynn Neary in 1982, had multiple iterations through the years. One involved Herb Howenstine hosting "Open Line Sports" at noon, followed by Tom Wiebell with a consumer advice call-in show, followed by the two-hour public affairs version of *Open Line*. In all, it was four hours of live talk programming every afternoon. Howenstine had developed his chops as a sports announcer for three years on WOSR, a low-power campus radio operation, before joining WOSU. Wiebell had a friendly, folksy personality and brought in a cadre of regular guests to talk about everything from computer tips to travel strategies.

Judy Cramer had taken over as host in the mid-1980s, followed by Alan Stavitsky. Howard Ornstein substituted for a time while searching for a permanent host. "People burned out quickly hosting the show," said Ornstein. "You were responsible for two hours a day with no support staff." Fred Andrle had produced a live call-in show on WOSU-TV and a variety of television programming in the '80s. He was taking a year off to travel across the country in his van when Ornstein called with an offer to fill in as a search began for a permanent host. Andrle returned to WOSU, got the job, and went on to host *Open Line* for twenty years, becoming the most popular news personality at the station.

FIG. 17.1. The cover of a 1994 *Airfare* featured the many hosts of *Open Line.* Left to right: Herb Howenstine with sports talk, Tom Wiebell with how-to talk, and Fred Andrle with current affairs talk.

Andrle's interviewing style was knowledgeable, civil, and easy on the listener. "My job is to call up the most interesting people in the world and talk to them," he said. "I get to learn what they are thinking, and most everyone, I have found, is available to us." Andrle was a poet, and his frequent experiments with the talk format included inviting audience members to call in to discuss their own poetry. Throughout his career he was honored often by local publications as the best radio talk-show host in Columbus.

IN THE KNOW

Twenty-eight-year-old Bill Schiffman was a surprise pick to host *In The Know* in 1983. The local CBS affiliate, WBNS, had carried the high school quiz series for years before deciding to offer it to WOSU. Schiffman, a CPA, had been a regular guest host during television pledge drives and claimed the only reason he was chosen was "because the crew liked working with me."

The Leo Yassenoff Foundation funded the initial seasons of *In The Know,* and American Electric Power has supported the series in recent years. Dale Heydlauff, president of the AEP Foundation, had served as president of the Friends of WOSU Board, and the series fit the foundation's educational mission. The original staff included producer Larry Wegman and quiz judge Carl Papai. In recent years Manuel Wilson has served as judge, with George Levert as producer.

Ohio State provides student scholarships to *In The Know* participants who reach the quarterfinals. Among those who led their teams to victory when the show was on WBNS was Richard Cordray of Grove City High School, who later become a five-time *Jeopardy!* champion and a prominent Ohio politician. In 2017 Cordray's son, Danny, almost single-handedly brought his team from Central Crossing High School, also in Grove City, to the championship.

"I'm very proud of this show," Schiffman said. "We've had some nine thousand kids on *In The Know* through the years, and this is their Friday-night football. We give them a platform where it's cool to be smart. That's the legacy of the show."

FIG. 17.2 Bill Schiffman started hosting *In the Know* in 1983. He would also chair the WOSU-TV auction, host the station's *Chefs in the City* fund-raising events, and serve as president of the Friends Board.

CLASSICAL 89.7

Through the late 1980s and 1990s WOSU's classical radio station continued to include an impressive team of on-air hosts and volunteers behind the scenes, including Anne Melvin, a prominent local philanthropist whose generosity would later prop up the struggling Columbus Symphony Orchestra. The station regularly broadcast concerts by the symphony and other local groups, and it aired operas from the Met as well as the popular national program *Adventures in Good Music* with Karl Haas. It also promoted local arts happenings. For several years in the 1990s and 2000s, Bill VanDervoort would bring into the station his extensive classical and opera record collection and spend a few hours raising money during the classical radio fund-raisers. VanDervoort's daughter recalled how they knew where their Dad would be every Saturday afternoon—listening to WOSU and the Metropolitan Opera in his den. He would turn up the volume on the family stereo so loud that the windows would be vibrating. His family finally gave him some headphones, which, his daughter said, may have led to some of his hearing issues. VanDervoort was one of many avid opera lovers who supported WOSU's classical station for decades.

FIG. 17.3. A 1980 "Road Rally" fund-raiser for WOSU radio including (left to right): Bill McNutt, Charles McCloud, John McGrody, Jody Williams, Mary Hoffman, Mike Fletcher, and someone unidentified.

With FM listenership growing, the classical station seemed in a good position. "Listener response is great," said classical music director Mary Hoffman at the time. "Folks travel around the world, and then come back to the area and say there's no other station like WOSU. That makes it all worth it." Boyce Lancaster held down the important early-morning shift, becoming the voice of WOSU through his work on station IDs, television promotions, and voice-overs.

Lancaster and music director Beverley Ervine were married in a ceremony at the station in 1986. There are numerous examples of matrimonial bliss propagated at WOSU. Mary Hoffman—then Mary Rousculp—had married newsman George Hoffman years earlier, and in 1984 chief television engineer Tom Lahr walked down the aisle with Marilyn Smith, also from the newsroom. Kevin Petrilla in radio operations later wed Cheryl Krebs from the marketing department, and the list goes on.

Classical evening host John "Jud" McGrody was hired by Hoffman in 1978 and developed a loyal following through the 1980s. Known for his acerbic wit and smooth delivery, McGrody was as great on the air as he was difficult off it, causing internal havoc for no apparent reason, Hoffman recalled. McGrody said he tried to develop a rapport with listeners, believing that many classical announcers "sound boring, stale and stodgy, leading people away from the greatest music in the world." He also wrote poetry and performed on Monday nights at Larry's Bar, a popular campus watering hole.

In 1986 McGrody was riding his bicycle to work when he caught a pedal on a curb and crashed, hitting his head on the pavement. He returned to the air several months later but admitted to the *Dispatch* that he had extreme mood swings and "there have been almost tidal waves of low self-esteem." He left for Chicago in 1991 with his wife, but he came back to WOSU every year to host the Columbus Symphony series. In 2002, as starkly written in his obituary, "Jud McGrody died in his Evanston home of a self-inflicted gunshot wound."

AM AND FM EXPANSION

Since the day in 1941 when WOSU agreed to accept the 820 AM daytime-only frequency, there was hope that someday the station would be on the air full-time. In 1988 it finally achieved that goal. The FCC had approved a WOSU proposal to broadcast beyond sunset at low power using a complex array of six towers in southern Columbus. The station could program all day and all night, although listeners north of the city would still have limited reception because of the reduction of power.

One of the successes of general manager Dale Ouzts, radio manager Sam Eiler, and the engineering team during this period was the expansion of FM coverage across Ohio. From 1989 to 1998 WOSU developed four new FM stations to carry the classical music format. The first was WOSV in Mansfield at 91.7, in coordination with Ohio State's regional campus there. WOSP in Portsmouth, along the Ohio River, was next at 91.5. WOSE in Coshocton at 91.1 was developed after strong support from local civic leaders, led with enthusiasm by David Baker, who later became president of the Friends Board. The group raised more than $60,000 to create an endowment to support station operations. At Ohio State's regional campus in Marion, leaders helped raise funds to plant a tower and transmitter in a soybean farm outside town. WOSB, which covers a 20-mile radius also at 91.1, was the last of the four new stations to hit the air.

The promise of developing local programming at the four stations never materialized, but the network helped WOSU extend the classical music signal and develop local relationships. Today, the Coshocton and Marion stations continue to provide funds from endowments to support operations. With the coming of digital high-definition radio, some of the stations provided a second channel of WOSU news programming for those listeners with HD radios.

KEITH AND MARGARET TYLER

WOSU lost a link to its history going back to the 1930s when eighty-nine-year-old Keith Tyler and his wife, ninety-four-year-old Margaret Tyler, died in an automobile accident in 1994. The Tylers were revered by many who knew of their influence on educational broadcasting.

Tyler was among the earliest researchers in educational radio, with studies dating to 1935, and he was most proud of being a key player in reserving channels for educational television in 1952.

In an oral history Tyler thoughtfully tackled the subject of ethics. Citing a former student who had recently been fired as a PBS manager, he said: "Sometimes you have to put your job on the line and simply say, 'This is what I stand for,' but nobody always gets everything he wants. You'd better think about what you are going to face and decide beforehand how you're going to behave in a situation. Then when you are caught up in it, you've thought it through. You know where your principles are, where the ethics are, and you behave that way."

A month after Tyler's death, WOSU announced it was eliminating the Ohio State Awards program he had started in 1936. The awards were the oldest in broadcasting, but it was a time of cost cutting, and attempts to gain corporate or university support for the program had failed.

THE HUBER REPORT

WOSU's fiscal situation continued to be tenuous into the 1990s and retired radio manager Don Davis, who chaired the station's Friends Board in 1994, initiated a meeting with Gordon Gee to express concern about the level of university funding. Despite Gee's flippant comment a few years earlier about eliminating WOSU, the president was vigorous in telling Davis and other board members that the university would never divest itself of the stations. Gee said he viewed the radio and television operations as valuable resources for the university and the community. After the meeting, Davis reported that Gee and Ed Ray, the senior vice provost, had agreed to focus more effort on resolving WOSU's fiscal issues.

There was no word from the administration in the coming year, but Randall Ripley, dean of the College of Social and Behavioral Sciences, met with the WOSU board and agreed to yet another study. Researched and written by retired dean Joan Huber and issued in 1996, the report compared WOSU with eleven other university-run stations. Huber found that WOSU's position at a major university in a midsized community was unusual. Only seven public stations in the top fifty markets were affiliated with universities. Market size usually drives the proportion of support provided by a university, with smaller stations requiring a much higher percentage of their budget from their institutions. For instance, Huber found that WOUB at Ohio University in Athens received more than 75 percent of its budget from the university.

Stations comparable to WOSU were difficult to find, leading Huber to cite a popular bromide: "When you've seen one public broadcasting station, you've seen one public broadcasting station." Despite that, Huber's research was revealing. Nearly every university licensee she surveyed was seeing declines in the proportion of funding from university and state sources, resulting in increased dependence on viewer, corporate, and earned revenues. This accounted for her description of WOSU board members, who felt they were "cartoon creatures, running faster and faster to stay in one place." Her analysis showed that university funding for WOSU had risen minimally since 1980, but that the budget had grown nevertheless due to more support from the community. From 1991 to 1996 the WOSU cash budget increased by $1.3 million, nearly all from the community. This naturally meant that WOSU began to gradually shift its programming to serve the broader central Ohio area, moving away from a university-centric focus. Huber supported continuing university support of WOSU but believed the current fiscal situation "will likely persist."

The WOSU board found the report positive, as did Ohio State administrators, but there was no immediate action related to additional university support.

THE AM DILEMMA

WOSU's commitment to news was evident during the riot and takeover by hundreds of inmates at the maximum-security Southern Ohio Correctional Facility in Lucasville in 1993. The News 820 staff provided live coverage during the eleven-day hostage crisis, which resulted in the deaths of nine inmates and one guard and caused more than $40 million in damage.

In the early 1990s the station also unveiled a daily morning newsmagazine with in-depth news, analysis, and arts coverage to go along with *Open Line,* the popular two-hour local talk show. The emphasis on local news was becoming common at WOSU, but supporting the coverage and the increasingly costly NPR programming was problematic.

A decline in listening to AM radio had started becoming noticeable across the country in the 1980s. By 1995 more than 75 percent of all listeners had moved to the FM band. Consultants assured WOSU management that the audience erosion wasn't the fault of the station's news format or talent. "It's AM. . . . The band is literally dying," they said. WOSU's classical music service had found a home on 89.7 FM when it went on the air in 1949 and took over the frequency full-time in 1980. WOSU-FM's weekly listening was double that of the expensive AM news operation.

Staying on AM showed little promise. WOSU's radio news service was stuck in a slow spiral of audience loss. From 1990 to 2000, the station's average weekly audience numbers fell about 40 percent. In the station's five-year plan, the paltry goal for WOSU-AM listenership was to be among the top twenty-five of the forty radio stations in the Columbus market.

A NEW CENTURY

WOSU's TV offerings at the turn of the century included the long-running high school quiz program *In The Know*; *Viewpoint,* a public affairs show hosted by Brad Mitchell; and *d'ART,* a cultural arts series. In the early 2000s the station produced several major documentaries, including one about Lustron homes, the prefabricated metal houses built after World War II by a Columbus-based firm. The team also produced two programs about the descendants of Thomas Jefferson and his slave Sally Hemings, hundreds of whom gathered at the Fawcett Center for a national broadcast about the controversy surrounding the relationship.

Classical 89.7 broadcast Columbus Symphony concerts throughout the decade, with the always entertaining Christopher Purdy as host. Purdy continued a tradition of WOSU announcers with in-depth knowledge and big personalities. In a profile in 2003 he recalled the advice of his voice teacher at Boston University: "You have no talent, so don't sing anymore. You do, however, have a wonderful mind and a big mouth, so go out and be intelligent." He was right, Purdy said: "I can't sing. I can't make music, but I'm very good at talking about music."

One of the most successful fund-raising events of the period was developed by WOSU staff, volunteers, and the Friends Board in 2001. It featured tours of the newly renovated Ohio Stadium, performances by the Ohio State Pep Bands, and remarks from head football coach Jim Tressel and athletics director Andy Geiger. About thirty-seven hundred patrons showed up for the event, which grossed more than $117,000 and won the PBS Development Award for special fund-raisers.

WOSU began to go online during this period, unveiling its own station website in 1998. By 2001 both the AM news operation and the FM classical station could be heard streaming throughout the world on wosu.org with a 32k-per-second signal. It was said to be the "best audio streaming you've heard on the internet."

It was a long time coming, but in 2001 WOSU's reporting line was moved from the College of Social and Behavioral Sciences to the vice president of university relations.

Within a few months, Dale Ouzts announced his retirement. Ouzts's brash style had brought mixed results during his twenty-three years as WOSU's general manager, but his southern wit and deep knowledge of public broadcasting had won him a number of local admirers, including many on the Friends Board. Ouzts traveled to China and Romania to mentor broadcasters in those countries and served as chair of the NPR board during a contentious fiscal period.

According to a *Dispatch* profile by reporter Julia Keller, who later won a Pulitzer Prize for feature writing, Ouzts was "endowed with an amazing ability to survive, to weather repeated storms that would have decimated less tenacious managers. He has endured a negative internal report that questioned his personal integrity; the unabashed enmity of some highly placed OSU administrators and faculty; and the nagging pressures of every station manager."

The consulting firm hired to assist Ohio State in finding a replacement for Ouzts was led by Ron Bornstein, who managed Wisconsin Public Radio for sixteen years and brought deep experience as a vice president within the University of Wisconsin system. Bornstein's team interviewed WOSU and university administrators, staff, and board members in 2002 to create an environmental scan for the final candidates for general manager. The scan identified WOSU's strengths, including the dedication and professionalism of the staff, the generally good quality of the programming, and the station's relative fiscal stability. The most obvious weaknesses cited by the team were the absence of an encompassing vision or strategic direction, a lack of community visibility, and a less-than-positive relationship with university administrators.

The search ended in August 2002 when Ohio State named the fifth leader in WOSU history: Thomas Rieland, who had directed the Center for Public Television and Radio at the University of Alabama for thirteen years, was hired as general manager.

STUDENT RADIO

Ohio State students periodically rallied for greater participation in WOSU, and each time the station would explain that it was not a student lab but a professional operation serving the entire state of Ohio. Bob Higgy had set the tone during the 1920s and '30s, believing that the station must always have a professional sound to compete with commercial stations. The tradition held with subsequent managers. The 1962 radio policy made clear the role of students: "The prime function of the station is not to train students, but rather to serve as an educational and cultural outlet." Some intrepid students developed their own form of radio, starting in the early years of the medium. Clair Tettemer worked with engineering students in 1940 to create the so-called Baker Broadcasting System. A studio was built in an empty storage room in Baker Hall, a men's dorm, and the station powered up to cover a radius of about three hundred feet, enough to reach radios in Baker and some nearby women's quarters. The station broadcast dorm news, student talent, and music. It went off the air when Tettemer left for military duty, but he revived the concept when he returned four years later. By then Baker Hall had become a women's dorm, and residents served as program directors, announcers, and marketing assistants ("We Cover Baker like a Blanket"). Things shut down for good after a few years when Tettemer left for a new job. He later surfaced as the first manager of the educational television station in Fargo, North Dakota, before returning to Ohio to take over the educational TV stations in Dayton and Oxford. In 1951 residents in Baker Hall and the dorms along Olentangy River Road began receiving a new student radio upstart, WOIO, sponsored by the Department of Speech. WOIO was a "wired wireless" station with telephone lines transmitting broadcasts from Derby Hall into the electrical outlets of the dorms. Listeners would plug in their radios and tune the dial to 1000 AM. WOIO lasted a decade before falling victim to money and technical issues.

A music-based student operation called WOSR started to broadcast to north campus dorms in 1971. Organized by Bill and Bob Rohrer, twins who were studying electrical engineering, the station was connected to seven dorms via phone lines. WOSR disappeared in 1976 after it attempted without success to get an FCC license for a 10-watt station.

A series of *Lantern* editorials in the early 1990s called for student radio time on WOSU following the demise of Ohio State's broadcast journalism department, but the station fought off any serious attempt at regular student programming. The Underground, an internet radio station, was created by students in 1995, and the university initially supported it with a $50,000 appropriation. The station also broadcast a single watt on 91.1 FM station called KBUX. Station leaders were caught illegally ramping up the signal in 1999. The operation survived a few more years before quietly disbanding.

Although Ohio State leaders often expressed support for student radio, they also had much larger issues to tackle. And while WOSU continued to employ students, it couldn't offer the in-depth experiences a student station provided. There was "absolutely no interest in the academic or the station side" in working together to develop student radio, said Thom McCain, a professor of communication who served on the WOSU Friends Board. "To run a student radio operation, you need adults, and no department was willing to be the adult," McCain said.

CHAPTER 18

PERSPECTIVE

2003–2018

Yesterday's strangest dreams are today's headlines and change is getting swifter every moment.

—LYNDON BAINES JOHNSON AT THE SIGNING OF THE PUBLIC BROADCASTING ACT, NOVEMBER 7, 1967

When LBJ signed the act that affirmed the future of public broadcasting in America, he looked beyond broadcast and predicted "a great network for knowledge—not just a broadcast system, but one that employs every means of sending and of storing information that the individual can rise." That would be the internet, the vast communications web that continues to change the way we send and receive print, audio, and video information. As LBJ said, "Change is getting swifter every moment."

This final chapter is an overview of the fifteen years between 2003 and 2018 at WOSU, which comes with a distinct bias because I've been involved in the challenges and accomplishments of the organization since I was hired as general manager.

I focus here on ten areas that have been most prominent during this period. Of course it's impossible to include in this limited space all the people and activities that have made WOSU such an important part of the central Ohio community, and I regret that many of them must go unrecognized.

Still, I hope these short essays capture some of what has defined WOSU Public Media in recent years.

A NEW PLAN

During my first few months at WOSU, I had my assistant, Paula May, set up individual interviews with all eighty-plus staff and many of the board members. It took a few months, but it was an interesting process and confirmed some of what the search firm's report had discovered about WOSU: There was a great deal of distrust in leadership and a lack of focus or understanding of WOSU's mission and goals.

We immediately initiated an extensive strategic planning process involving a core team of engaged and experienced staff, which purposely didn't include any of the senior leaders of that time. Within a six-month timeline, the team developed relational and core values for WOSU staff, a new mission statement, and a SWOT (strengths, weaknesses, opportunities, threats) analysis, and it identified initiatives and actions for five lines of business.

The planning process brought together staff members who had never even talked with one another to help build WOSU's future. The success of the process was important to my first years at the stations, and I thank the team members for their enthusiasm and dedication. Several of those core players were still an important part of WOSU more than fifteen years later, including Stacia Hentz, Christopher Purdy, Marcelita Haskins, Cindy Gaillard, and Mary Alice Akins.

One of the issues I heard from the interviews was that WOSU was an insular organization and needed to better understand where public broadcasting was going and what the best practices were elsewhere. We invited public broadcast leaders and other media thinkers to Columbus for "big picture" presentations to expand our awareness of the road ahead. Among them were public broadcast veterans Skip Hinton and Jim Fellows. Hinton talked about the importance of developing core beliefs based on "what you as a station value the most." Fellows saw great value in collaborations to create content and engagement in the community. Staff members were touched by a quote he provided from George Bernard Shaw that pointed to the important work they were doing: "This is the true joy in life, the being used for a purpose recognized by yourself as a mighty one."

WOSU needed new blood in key positions to expand its horizons, and in my first years I was able to hire not only a new television executive producer to take charge of local programming but also a new radio manager. Brent Davis is as close to a renaissance man as anyone I know. He brought to WOSU many talents as a creative writer (and published novelist), producer, musician, and even radio on-air host. He assembled a top-notch producing team and was adept at managing in a supportive, positive manner. Tim Eby brought a wealth of ideas to his role as radio manager, including an emphasis on building a stronger social media and online presence. As chair of NPR's board of directors, Eby was able to expand our knowledge about the larger issues facing public radio and compare our efforts to the best stations in the country.

During these first years at WOSU, I found some talented staff well prepared to move up within the organization. Susan Meyer took over marketing after serving as editor of *Airfare* magazine. Rob Walker moved into the lead position running WOSU membership. Mike Thompson advanced to news director. Mary Alice Akins was promoted to supervise the business operations after longtime staffer Sheri Walker retired. Through the years Walker had helped resolve many internal issues with her calm, wise counsel. Akins stepped into that important role. Her advice on issues great and small helped WOSU develop a healthy internal culture.

WOSU@COSI

During my family's first winter in Columbus, our two young girls were excited to escape the cold by visiting COSI, the Center of Science and Industry in downtown Columbus. COSI had relocated from East Broad Street to a massive new facility on the Scioto River a few years earlier. When I saw that some cavernous spaces seemed underused, a thought occurred to me: Moving our television studios from the basement of the Fawcett Center into a visible and popular destination might have some potential if it also helped resolve some of the issues COSI was struggling with as it adapted to its new home.

I sent a note to that effect to Kathryn Sullivan, the former astronaut (and first American woman to walk in space) who had headed

FIG. 18.1. David Chesebrough, COSI CEO, and Tom Rieland, WOSU General Manager, cut the ribbon in 2006 to open the WOSU studios and PBS-related children's exhibit space at the COSI Science Center.

COSI since 1996. She and Kim Whaley, COSI's chief operating officer, met with me and encouraged this unique collaboration. We focused on a boxlike space two stories high on the first floor that was used for occasional visiting exhibits. Over the next three years the WOSU@COSI studios took form.

As a newbie in Columbus, I had entrée to many who wanted WOSU to succeed but who also wanted to be inspired by a big idea. One piece of luck was that Barbara Snyder was serving as Ohio State's interim vice president of university relations during this period. Snyder was thrilled with the idea of extending WOSU resources into the community and convinced President Karen Holbrook to provide seed funding to develop concept designs and a business plan. Carl Kohrt, chief executive officer of the research giant Battelle, committed more than $1 million on behalf of Battelle. Other gifts helped us build the facility and buy some of the gear required to produce high-definition broadcasts from the space.

On September 28, 2006, WOSU and COSI unveiled the new digital media center. WOSU@COSI included television studios, video editing suites, a hands-on exhibit space for youngsters highlighting PBS Kids characters, and a media lab for teens. David Chesebrough,

FIG. 18.2. In one example of the many public forums at the WOSU@COSI studios, NPR's Neal Conan hosted a discussion about the intersection of science and religion.

who had recently taken over as COSI's CEO, cut the giant ribbon with me, and we celebrated with assistance from Caroll Spinney, otherwise known as Big Bird and Oscar the Grouch on *Sesame Street.* Two NPR reporters who had worked at WOSU, David Kestenbaum and Anne Gudenkauf, also showed up for the party, along with David Fanning from PBS *Frontline* and The Ohio State Marching Band.

With a presence downtown, WOSU became more relevant to the greater community, and the frustrating perception of it being simply a "university campus station" started to fall away. The WOSU public space at COSI hosted debates, forums, and performances and attracted a national public television conference in 2007. It hosted the president of PBS, Paula Kerger, who called the collaboration an "innovative model" for all public stations. A WOSU annual fundraising event, *Chefs in the City,* hosted by Bill Schiffman, brought together top local chefs and PBS chefs such as Rick Bayless, Hubert Keller, and Jacques Pepin in a series of special events in the WOSU@COSI studios.

There were challenges. The recession of 2008 hit as we were planning a second phase at COSI to build out space and create a radio studio presence. That never occurred. Although we raised $4 million for the project, WOSU had to take out a university loan to cover some renovation expenses. That loan was in addition to a previous university loan to build out long overdue new radio studios in the basement of the Fawcett Center. Like many other nonprofits, we struggled through the recession but recovered strongly, and WOSU@COSI became a decade-long success story for both organizations.

DIGITAL TELEVISION

After hundreds of millions of dollars in research development funds and twenty years of planning, digital television (DTV) was becoming a reality across America in the early 2000s. A federal mandate to go digital meant that valuable spectrum used by television stations would be opened up for sale to wireless providers such as AT&T and Sprint. For local stations it meant that new transmission and production equipment would be required.

WOSU started a DTV campaign in the late 1990s to bring attention to the need for an estimated $15 to $20 million for WOSU to make the transition. In the end the cost was less, and the bulk of the money came from public funding. WOSU scored federal grants worth more than $2.5 million to help with the transition. With guidance from state public television lobbyist Bill Blair, state capital funds helped provide the match necessary for the federal equipment grants. The state and federal funding along with some private monies provided enough support to build digital transmission and master control systems and to buy new digital studio and field production equipment.

The transition to digital provided not only stunning high-definition images but the ability to slice up part of the bandwidth to provide several additional standard-definition channels of programming to go along with the primary HD channel. WOSU began digital TV transmissions in May 2003 and started multicasting more than one channel, becoming the first station in Columbus to do so. WOSU broadcast its primary PBS channel in high definition and offered WOSU Plus with many how-to and cooking programs and WOSU Ohio, a network coming from Dan Shellenberger's governmental telecommunications shop at the Ohio Statehouse. Programming on the Ohio Channel ranged from legislative coverage and the annual state of the state from the governor to programming contributed from other Ohio stations. WOSU's two-hour daily radio talk studio was outfitted with robotic cameras to capture the program for the WOSU Ohio Channel. In 2017 WOSU added a twenty-four-hour PBS Kids channel to the lineup.

The challenge of transmitting digital pictures was just part of the equation. Viewers without a new digital TV set would have to use a converter box to receive the digital signals. In 2008, a year before the FCC mandate would move all stations away from the old analog broadcasting, an estimated ten million homes nationwide were "completely unready" for the conversion. To assist the many elderly and poor who were at risk of losing reception, the federal government made $40 coupons available to each household to help with the purchase of converter boxes.

WOSU saw this as an opportunity to serve local viewers who were at risk. Through the *Connect the Community* project, we encouraged all citizens to apply for coupons whether they needed them or not and donate them to the station. WOSU would use these coupons and additional support from the Columbus Foundation to purchase converter boxes and distribute them through its partners on the project: LifeCare Alliance, the Central Ohio Area Agency on Aging, and several Latino and Somali organizations. The idea was hatched at a Friends Board meeting after a suggestion from Bill Schiffman.

We established a hotline for viewers and had a series of call-in programs. One of our partner agencies reported more than five hundred requests for help from the elderly during a single week. Staff members and volunteers gathered on several Saturday mornings to install converters in homes and apartments where they were needed most in Columbus.

Chief engineer Tom Lahr called the effort the most gratifying project in his four-decade career at WOSU. Lahr was already well known for handling calls about TV reception issues by showing up at viewers' homes to adjust the antenna. The work of installing converter boxes was just an extension of something he had been doing for years.

WOSU received a national Community Impact Award for Engagement in 2009 from the Corporation for Public Broadcasting, which recognized the station for its extensive outreach and innova-

FIG. 18.3. Tom Lahr was joined by staff and volunteers to install converter boxes in homes and apartments across central Ohio to allow older televisions to receive the new digital television signals.

tive model for helping those at risk. WOSU in turn presented Lahr, Dave Carwile, and marketing associate Carl Higginbotham with the Higgy-Tyler Staff Award for their leadership on the project.

Though not mandated by the government, radio stations across the country were also making a transition to digital. WOSU 89.7 became a digital HD Radio station on April 5, 2005, the first in Columbus. The new technology allowed stations to split their signal into multiple crystal-clear channels, but listeners needed a special radio to receive the signals. HD Radio was not a mandatory transition like digital television, and adoption has been slow. Today, more new cars are being built with HD Radio as a standard feature, but the many online and other audio choices have left the channels struggling to attract listeners.

ALL SIDES WITH ANN FISHER

In May 2009 Fred Andrle hosted his final *Open Line* radio broadcast. After twenty years of interviewing some of the most influential people in the world, Andrle retired to write poetry and seek new adventures. WOSU news director Mike Thompson saw Andrle as the "true definition of a fair and balanced host" who made the job seem easy while it was in fact one of the most difficult in radio. I was quoted in the *Dispatch* as saying: "Fred has been a special gift for Columbus for the past two decades." *Dispatch* columnist Joe Blundo wrote a poem dedicated to Andrle that compared the civility of *Open Line* to the raucous talk shows on AM commercial radio. Here is a portion:

> 'Twixt 13-minute commercial break,
> A maelstrom of BS and bile
> From end to end on the radio dial,
> And where find shelter from the dread?
> In the one safe harbor, name of Fred.
> Tender eardrums did he spare
> From bombast bursting on the air.

Fred Andrle was a tough act to follow. It took several months of discussion and tryouts by potential hosts before WOSU decided to transition to a new show. On September 8, 2009, Ann Fisher, an editorial writer and columnist for the *Dispatch,* started hosting *All Sides with Ann Fisher.* Fisher brought journalistic depth, great local contacts, and creative energy to the position, but listeners had to adjust to an on-air style quite different from that of the smooth Andrle. Having never been a broadcaster, Fisher grew into the job literally on the air. She soon developed a strong following and become a popular guest host and interviewer at major community events. The show became more news-oriented, with segments shortened to allow multiple topics every hour, and it continued to be streamed online and could be seen live on WOSU Ohio, one of the digital channels of WOSU-TV.

During major breaking local news, *All Sides* became the place to go for listeners. Such was the case the morning of November 28, 2016, when a text arrived: "OSU Safety Alert—Active shooter on campus, shelter in place." WOSU immediately started its coverage. Reporters Esther Honig and Sam Hendren rushed to the center of campus, and Mike Thompson joined Fisher on *All Sides* for live

FIG. 18.4. Fred Andrle was replaced by Ann Fisher after his retirement from *Open Line* in 2009. The new radio talk show was called *All Sides with Ann Fisher.*

coverage. The digital team managed a constant stream of updates on wosu.org, Twitter, and Facebook. Honig and digital producer Gabe Rosenberg filed reports for NPR and the BBC.

As it happened, the situation did not involve an active shooter after all. A driver had rammed his car into a group of students and faculty near Watts Hall before exiting and charging at them with a knife. An Ohio State police officer shot and killed the assailant, a Somali student, just two minutes after the attack began. Eleven people were hospitalized.

WOSU television, radio, and digital all covered the multiple news conferences that day. In the following weeks *All Sides* devoted shows to follow-up coverage with Ohio State president Michael Drake, members of the Muslim and Somali communities, and experts on rampage attackers, ISIS, and post-traumatic stress.

The year before the attack, WOSU-TV had produced an online video called "Surviving an Active Shooter" for the university's Department of Public Safety. The video had been viewed more than one million times and used in training across the country.

THE BIG SWITCH

In 2006 WOSU attempted to purchase the FM station of a local religious broadcaster, but after six months of negotiations the deal fell through. That sent us back to the drawing board. Research showed that the WOSU-FM classical audience had decreased by 25 percent in a ten-year period, while nationally the public radio audience had grown by 35 percent. Coupled with the deficit spending on WOSU-AM's news operation and the consistent loss of AM listeners, something had to give. In January 2008 WOSU brought morning and afternoon NPR news from the AM station to WOSU-FM and the four repeater stations for the first time since the 1970s. The FM station was now a mixed news/classical format.

Classical music continued on 89.7 FM between 9 a.m. and 4 p.m. and in the evenings. The station promoted the airing of classical music programming twenty-four hours a day on the WOSU-FM HD-2 channel and online, but that was little consolation to many classical fans who couldn't understand why WOSU programmed the same news on both its AM and FM stations.

The year 2009 marked a transitional time in leadership. Ed Clay retired as television station manager, and Tim Eby, who as radio manager had driven the move of more news programming onto 89.7, left to run St. Louis Public Radio. I decided to reorganize by hiring public broadcast veteran Karen Olstad as chief operating officer to lead collaboration among all content and technical areas. Olstad's six years as COO marked one of the most transformational periods in our history, aided by her selfless, thoughtful decision making.

Karen and I became very focused on a strategy to help resolve the long-standing AM dilemma that hung like an anchor around the station's future. One option was to move the news programming entirely to the 89.7 FM station and shut down the classical station or move the music online. Other public stations had used that strategy and survived the listener onslaught. We decided to go in another direction.

After months of negotiation and with support from leadership at Ohio State, especially Senior Vice President Tom Katzenmeyer, WOSU was approved to purchase the 101.1 FM frequency in Columbus. The station owner was ready to cash out, and in Decem-

FIG. 18.5. Philanthropist Ann Melvin (center) helped make the purchase of Classical 101 possible in 2010. Ann is joined by the Classical 101 team (left to right): Boyce Lancaster, Jennifer Hambrick, John Rittmeyer, Beverley Ervine, and Christopher Purdy.

ber 2010 it became Classical 101 with an all-day classical music format. Major funding for the purchase came from long-time supporter Anne Melvin and a few others. Melvin had volunteered and then became a part-time staffer working in the music library at WOSU in the 1980s.

On January 6, 2011, WOSU-FM took to the air for the first time as an all-news station with NPR news, local news directed by Mike Thompson, and local talk anchored by Ann Fisher's two-hour talk show, *All Sides*. After broadcasting on the frequency since 1941, WOSU sold the historic 820 AM station the following year. WOSU radio in Columbus now consisted of two FM radio stations serving up all-day news and classical music services.

An analysis by the nonprofit consultant Public Radio Capital clearly outlined the importance of moving to FM, but the move was not without controversy. The 101.1 FM signal covered a smaller area, and some listeners in parts of Columbus couldn't tune in without interference. WOSU placed the classical signal on the HD-2 digital channel of 89.7 for those with HD radios and also online to ease the reception issues, but some fans were still displeased. My position

FIG. 18.6. In 2011, Ken Burns and Lynn Novick were guests of WOSU for a sold-out event at the Mershon Center to discuss their Prohibition documentary. Longtime supporters Mary and Bob Lazarus with Burns at a WOSU speakeasy party.

was that classical music and opera would always have a devoted niche audience, but our strongest signal had to be devoted to news programming. I also pointed to communities such as Milwaukee and Kansas City that had seen their local classical radio stations disappear. That would not occur in Columbus.

The consultants had predicted that once WOSU switched its news product from AM to FM, the audience would grow slowly for years to come as people found the station and gradually changed their listening habits. It was true. In the first five years of 89.7 NPR News, the audience grew an average of more than 15 percent annually. By 2017 the station was often among the top five morning-drive stations in Columbus, and 89.7 NPR News was regularly named the best radio news operation in the state by the Ohio Society of Professional Journalists and the Associated Press.

COLUMBUS NEIGHBORHOODS

It was nearly midnight, and I was walking along High Street toward Goodale Park in the area near downtown known as the Short North. I had just left the neighborhood screening of a WOSU documentary that told the story of the transformation of the Short North from a rough, crime-ridden area to a vibrant arts community.

As I walked through the neighborhood that evening, I was so proud of our team. Longtime residents of the neighborhood had expressed amazing appreciation after the screening, saying the documentary gave them a greater sense of place and that no one would ever tell their story like we had. It was one of the most gratifying evenings in my career.

The Short North program was the first of a dozen historical and cultural documentaries produced by WOSU from 2010 to 2016 that focused on the distinctive histories and characters of the city's many neighborhoods. Though I conceived the idea for the *Columbus Neighborhoods* series as a celebration of the city's bicentennial, it picked up steam under the creative leadership of Brent Davis, Cindy Gaillard, Mary Rathke, Diana Bergemann, George Levert, Ben Bays, and other talented folks. Major funding came from American Electric Power, J. P. Morgan Chase, and State Auto Insurance; other organizations and families across the community also supported the effort.

The project was the first to build an extensive digital presence and to include storytelling sessions at places such as a Baptist church in the King-Lincoln district and art galleries in the Short North. It was always intended that segments from the documentaries would live on in local schools. The *Columbus Neighborhoods in the Classroom* project, directed by Marcelita Haskins, won a national educational award for its video learning modules and lesson plans culled from the documentaries for use in local schools.

The Short North documentary was followed by others highlighting German Village; King-Lincoln; the University District; Downtown and Franklinton; Olde Towne East; the Southside; Worthington; Bexley; Clintonville; and the Tri-Villages of Grandview Heights, Upper Arlington, and Marble Cliff. The series concluded with a history of the new Americans, the immigrants who continue to settle in Columbus and make the community a fascinating place. It won an Emmy Award.

Columbus Neighborhoods also won the coveted Artistic Excellence Award from the Greater Columbus Arts Council, along with many local honors and regional awards, and reruns even today continue to draw good viewership.

FIG. 18.7. WOSU's Mary Alice Akins, Alysia Gobert-Smith, and Sheri Walker hosting a *Columbus Neighborhoods* event at the Lincoln Theatre in 2011.

FIG. 18.8. In 2014, WOSU won a raft of regional Emmy awards for local television programming, including a documentary on the life of Senator John Glenn and a local arts program, *Broad & High.*

The documentaries challenged our production and videography team, and they were more than up to the task. The series became a weekly half-hour show in the fall of 2016 with shorter profiles of neighborhoods across central Ohio.

From 2012 to 2015 WOSU's documentary work also included three popular projects produced by Ben Bays and Shawn Likley. The first was *Jesse Owens: Enduring Spirit,* a profile of the Olympian and Ohio State student, followed by *John Glenn: A Life of Service,* celebrating the astronaut and statesman. Both won regional Emmy awards. The other major project, *Admiral Byrd: Forged on Ice,* focused on explorer Richard Byrd, namesake of Ohio State's acclaimed polar research center. All of the documentaries, produced with some financial assistance from Ohio State's Department of Athletics, connected WOSU to the university in unique ways and were picked up for distribution on PBS stations across the country.

ARTS AND CULTURE

From its birth in the 1920s, WOSU has played an important role in focusing attention on local artists, artworks, performances, and exhibits. Because all radio was live and local in its first decades, the station regularly invited musical groups into the studio to perform and even developed its own symphony and drama troupe.

Today, Classical 101 provides all-day access to classical music with local hosts who profile guest artists and maestros and broadcast local performances by the Columbus Symphony and ProMusica, and from Ohio State. The 89.7 NPR station retains its pure news format throughout the week, but on weekend nights it airs *Bluegrass Ramble,* continuing the long tradition of that music genre on WOSU.

Television station manager Ed Clay was a talented jazz musician, so it's not surprising that a variety of local jazz programs showed up on WOSU-TV during his three-decade tenure. In one example, *Just Jazz* premiered as a series in 1989 with local groups, including Gene Walker's *Generations and Actual Proof* with Jeanette Williams performing before a studio audience. The producers also worked with Columbus performance artist Robert Post to tape a solo work of

FIG. 18.9. A promotional advertisement for WOSU-TV's local programming in 2018.

comedy, theatre, and movement arts. *Robert Post—In Performance* went on to win a regional Emmy award.

Under the title of *d'ART,* WOSU produced short, well-regarded documentaries about the local arts scene for several years beginning in the late 1980s. They included features on ceramicist Christopher Gaston and neon artist Corbett Reynolds. When I came to town, one of the first people to host a welcome party was Ray Hanley, the provocative director of the Greater Columbus Arts Council. He wanted to elevate the arts in Columbus and thought WOSU could help. Ray and I forged a unique collaboration to begin a monthly TV arts program and short video segments called *ArtZine.* The series, produced by Cindy Gaillard, morphed into a weekly program called *Broad & High,* with Jackie Shafer producing literally hundreds of profiles of local artists in central Ohio. Tom Katzenmeyer, who runs the arts council and was once my boss at Ohio State, expanded the collaboration between our two organizations.

In 2012 WOSU celebrated the city's bicentennial by coproducing *Jazz Moves Columbus,* a live performance of the Columbus Jazz Orchestra and BalletMet that played to full houses at the Riffe Center for two weeks. WOSU also partnered with Byron Stripling and the jazz orchestra for a live-streaming broadcast of special performances seen online throughout the world. WOSU's Christopher Purdy hosted and our TV production team assisted mightily in the success of an unprecedented collaborative arts performance called *Twisted* in 2014 and again in 2016.

Coverage of the vibrant local arts and culture scene in central Ohio is a distinguishing characteristic of WOSU Public Media and continues to be one of the key pillars of the organization.

DIGITAL MEDIA EXPLODES

For more than twenty-five years, WOSU's brand was WOSU Stations. In 2006 we strategically moved away from the focus on "stations" to "WOSU Public Media" as way to position the organization as a multimedia provider of content for broadcast and the growing digital world. WOSU launched its mobile app in 2013 so listeners of the news and classical radio stations could tune in with their smartphones or tablets. The app was downloaded more than fifty-four hundred times in its first year. The biggest digital hit that year was the story of teacher Carla Hale, fired after nineteen years at Bishop Watterson High School when it was discovered she was gay. The coverage received more than sixty thousand page views on WOSU's website. It was a record at the time, but it's been exceeded many times since.

As smartphones proliferated, so did the viewing of video on portable devices. Blogs, podcasts, apps, and video and audio streams became the expected backbone of any broadcast station. A social media presence was also mandatory to provide links to content and reach new audiences. WOSU added digital expertise to its news and cultural arts programming, and it started to pay off immediately. Video views of segments from programming often hit gold online.

"The Man Who Knits," about a retiree in Westerville, had more than one million views; a piece about a vegetarian soul food business in Columbus hit similar numbers.

Public and commercial media are still struggling with the ever-changing world of digital distribution and how to build a revenue model around it, but interest in the content WOSU produces through digital and social media is only going to grow, and WOSU has to be where the audience is going.

A GROWING EDUCATIONAL ROLE

Education has always been part of the DNA of stations founded within universities. WOSU has focused on serving the general public along with K-12 and college students and teachers since its beginnings in the 1920s. Its strongest initiatives included the Ohio School of the Air, and projects such as *Ready to Learn,* which was launched in 1996 and focuses on early childhood learning through training for parents, preschool caregivers, and teachers. State funding and grants have helped sustain the highly successful *Ready to Learn* program through the years, with Marcelita Haskins and Jan Pollard leading this work at WOSU.

The Bakery Shop, an online game that teaches simple economics to youngsters in early elementary grades, has been one of WOSU's most successful educational offerings. Created in 2009 through a collaboration with Ohio State's Advanced Computing Center for the Arts and Design, *The Bakery Shop* remained popular on WOSU's website some eight years later, an eternity when it comes to digital games. In 2017 it was accessed nearly fifty thousand times.

WOSU's local efforts were aided by a new emphasis on building educational resources from PBS, which developed PBS Learning Media, a digital archive of thousands of educational resources for teachers. Over five thousand teachers and parents in central Ohio were registered for Learning Media in 2018. PBS also unveiled a new twenty-four-hour Kids Channel that was picked up by most affiliates, including WOSU-TV.

The big news in 2017 was the addition of a new educational division for WOSU, the roots of which date back to the earliest days of the television station. In 1957 Columbus Public Schools partnered with WOSU to produce and air telecasts for in-school viewing. Five years later a nonprofit called the Central Ohio Educational Television Foundation was formed. That organization would change its name to Instructional Technology Services of Central Ohio (ITSCO) to continue working with teachers on using media in the classroom and providing technology training.

In 2017, after extensive discussions, ITSCO director Amy Palermo dissolved the organization as a nonprofit and merged with WOSU Public Media as we created a strong, new division to serve teachers and students called WOSU Classroom.

FRIENDS OF WOSU

The volunteer Friends of WOSU Board is a valuable advisory group of community and campus supporters of WOSU who meet to discuss the major issues facing the stations and advocate for the organization. In preparation for the difficult budget years of the late 1970s and early '80s, the board once spent eight hours with staff detailing every line in the WOSU budget. Every contribution was to the Friends of WOSU during that time, and the board paid extremely close attention to how WOSU spent those funds. In the years since, the Friends Board has been less focused on the budget, as the funds are now directed to WOSU as an organization. The board advises on major projects, programming, membership, and major giving strategies; it also promotes volunteering in numerous ways. Volunteer assistance beyond the board had been a staple of WOSU since its earliest days. Bob Yocom has volunteered for many computing and membership tasks for over thirty years starting in 1986. He is one of many.

A list of past board presidents is included at the end of the book, but I've been lucky to have been mentored about Columbus from some of the community's leading citizens. Members have brought a fresh perspective to many of our activities, from major giving strategies to digital marketing to new program concepts.

The Friends Board began presenting the coveted I. Keith Tyler Staff Award in 1997. Nominations came from staff, volunteers, and board members. In 2005 I asked the board to add the name of Robert C. Higgy to the award to reflect his lifetime of devotion to WOSU. Higgy and Tyler had a strained relationship and may not

FIG. 18.10. The WOSU Friends Board awarded the coveted Higgy-Tyler Staff Award at the 2017 WOSU holiday party to producer Diana Bergemann. See "WOSU Staff and Board" at the end of the book for a list of all past WOSU staff honorees.

have appreciated being joined in such a manner, but the award now honors both for their contributions to the success of WOSU and educational broadcasting.

Today the board chooses the honorees and provides a financial reward for each winner. The presentation of the Higgy-Tyler Award is one of the highlights of our annual holiday party.

FIG. E.1. WOSU staff ham it up on the *In The Know* set in 2005, with the iconic O-H-I-O.

EPILOGUE

WOSU AND OSU

I've always believed that building strong ties between WOSU and its license holder Ohio State was of prime importance. Yet, that goal is like walking a tightrope. WOSU must show its worth to the campus and abide by the many university rules and processes, while serving over two million citizens with programming important to their lives. Generally, about 85 to 90 percent of annual funding comes from outside the university, most from individual memberships, corporate program support, and earned revenue. Still, there must be an allegiance to the license holder, and when WOSU is perceived as moving too far away from its university core, trouble ensues. One thing you learn about studying WOSU's history is that it's nearly impossible to accomplish anything of significance without the support of the university. Building healthy relationships with university administrators remains critically important for WOSU leadership.

I often tell Ohio State leaders that WOSU is the most effective representation of the university land-grant mission. The university seal includes the motto *Disciplina in civatem,* translated from Latin to *Education for Citizenship,* reflecting the greater role of the university in educating all the citizens of Ohio. That is a role that WOSU owns like no other university entity.

In 2014, WOSU developed its first-ever set of editorial integrity guidelines, and it has been used as a national model for other university stations. We believe that the university is best served by allowing WOSU a free and independent decision-making process, holding WOSU ultimately accountable to the public and not the license holder. In short, the university shouldn't dictate programming or interfere in news coverage. Among those who signed off on the integrity standards was senior leadership at Ohio State.

In my first thirteen years at Ohio State, I had six different supervisors. Such was the rate of burnout and turnover in the university communications head job. In 2015 I was given the rare opportunity to move WOSU to the Office of Administration and Planning. Jay Kasey's broad range of supervisory influence as senior vice president of that area included OSU facilities, public safety, printing, and transportation. Kasey's leadership team was focused on building quality systems, while seeking fiscal efficiencies. WOSU quickly adopted new models of operating, which included transitioning IT, human resources, and some finance areas to centralized hubs. Our philosophy was to create as many behind-the-scenes efficiencies as possible to focus more resources on local programming.

The move to this new division had an even greater impact when a new collaboration between COSI and the American Museum of Natural History meant that the WOSU studios at COSI would be no more. We were stunned to hear we were going to be displaced by a dinosaur exhibit and lose this studio facility touted nationally as unique in the country. Relocating production equipment and staff back to the basement of the Fawcett Center was difficult, but being booted out of COSI in 2016 turned into a blessing.

Kasey made a suggestion that we consider building a similar studio as part of Ohio State's planned arts district at High Street and 15th Avenue. Karen Olstad, chief operating officer, encouraged a bigger vision: moving all of WOSU to the new development and building a new headquarters. Meetings with Keith Myers, who supervised real estate for the university, and Amanda Hoffsis, who led the Campus Partners development bordering campus, quickly led to a concept of bringing all of WOSU into the new district on a footprint at 14th Street and Pearl Alley.

We had support from the administration. All we needed was money. The partnership with COSI had been the largest capital project in our history, but this would dwarf it. Most estimates put the cost at up to $30 million for the building, furnishings, and equipment. With the luck of timing, the FCC spectrum auction occurred in 2017, and WOSU and Ohio State leaders agreed to put WPBO-TV in Portsmouth up for bid. Several PBS stations already covered that region of southern Ohio with a TV signal. The unusual FCC reverse auction process ended with a payment of $8.8 million to Ohio State for a noncommercial television operation worth about one-tenth of that amount. With Kasey's advocacy, university president Michael Drake agreed to set aside $7.8 million for the new building and equipment, while a $1 million endowment was set up to fund student experiences within WOSU. It was a start, but we had to make a quick dent in our $12 million philanthropy goal or the project was dead.

I had known Andy and Sandy Ross for many years when our Advancement Director, Laura Baker, and I strategized about asking them to become principal donors for our new headquarters. They had been helping us for decades with challenge gifts during pledge drive. Neither of us had ever asked someone for a multimillion-dollar gift, but we were strangely confident when we met at their home on a sunny March day in 2017. We knew that their love for all of WOSU's programming and services was incredibly strong. Andy and Sandy had seen the impressive artistic renderings of the new WOSU facility, and they listened intently as we described the vision for our presence at the nexus of campus and community, including the unique community engagement space and media lab that was part of our plans to build public media for the future.

When we made our $5 million request for support, there were a few seconds of silence and then Andy, in his folksy way, said, "Well, that might be possible." Sandy agreed, flashing her bright smile. We stood for a group hug, all four of us. It may have looked like the famous final scene of *The Mary Tyler Moore Show.* They were as excited as we were.

That was the day I believed that the new WOSU Public Media headquarters, built to serve central Ohio citizens for another one hundred years, would become a reality.

FIG. T.1. Tom and Carol Rieland (right) talking with WOSU donors Andy and Sandy Ross at the campaign kickoff to raise funds for the new WOSU headquarters in October 2018.

FIG. T.2. A reunion of some of the oldest alumni of WOSU in 2017 with WOSU general manager Tom Rieland (center). Left to right: Gene Gilliom, John Schmidt, David Ayers, and John Morison.

HISTORICAL TIMELINE

1895: Guglielmo Marconi's experiments in wireless telegraphy, conducted when he was twenty-one, resulted in the transmission and reception of a signal over a distance of about 2 kilometers from his family's villa near Bologna, Italy.

1899: "The Tesla Club" is started by juniors and seniors in Ohio State's electrical engineering program to share current papers and news about the visionary work of Nikola Tesla, including his research into wireless technology.

1900: The first documented campus experiments in wireless telegraphy at Ohio State are performed by electrical engineering student Robert Marriott. He became nationally recognized as a wireless engineer and founded The Wireless Institute.

1909: The first course in wireless telegraphy is offered at Ohio State due to Robert Marriott's contribution of an array of wireless transmission equipment.

1913: Ohio State's wireless station plays a key role in communicating to the rest of the country about the Great Flood in Columbus that claimed ninety lives.

1917: The university wireless station 8YO is shut down under presidential order as WWI begins.

1918: The first wireless voice transmissions in Ohio are sent across the Olentangy River as part of an Army radio class working with the faculty of OSU electrical engineering.

1920 (April 20): The first regular voice and music radio programs in Columbus are broadcast using the 8XI experimental station at OSU. Transmissions later that year included daily weather and market reports for Ohio farmers. This marks the anniversary date of regular radio broadcasting at The Ohio State University.

1922 (June 3): Ohio State is assigned the call letters WEAO, becoming the sixth educational radio station licensed in the country.

1923–24: During the academic year, WEAO broadcast one hundred seventy-four lectures, daily market and weather reports for farmers, a variety of music programs, and OSU football games.

1926: Under the leadership of engineering faculty member Dr. Charles Wright, two antenna towers are built at Woodruff and Neil avenues and are connected to a new 750-watt transmitter located in a wooden building between the towers.

1927: Twenty-six-year-old Robert Higgy becomes the first full-time director of the radio station after Dr. Charles Wright retires due to health issues.

1929: The Ohio School of the Air begins broadcasting from the WEAO studios, with transmissions rebroadcast over powerful WLW in Cincinnati to classrooms across the Midwest. Other states would soon follow and develop their own schools of the air.

1930: The first annual Institute for Education by Radio was sponsored by Ohio State as leaders in educational and commercial broadcasting spent a week discussing the use of radio for public service and education.

1932: A financial report for WEAO showed an annual operating budget of $17,531.

1933 (September 1): Ohio State is granted the new call letters of WOSU for its radio station to reflect the initials of its license owner.

1934: The Radio Junior College is developed to offer radio courses in business, French, and a variety of other subjects during the Depression years. Over ninety-three hundred people enrolled for one or more courses across Ohio in the college's first three years of existence.

1938: The radio tower and transmitter for WOSU are moved to the University Golf Course. Station supervision moves from the College of Engineering to the President's Office.

1941: WOSU AM finds a permanent frequency at 820 AM after many moves around the dial and begins transmitting at 5,000 watts daytime only.

1944: Ohio State president Howard Bevis testifies before the Federal Communications Commission in support of educational-license set-asides for the new FM radio band.

1949 (December 13): WOSU-FM signs on at 89.7 within the new educational portion of the FM band. The station initially duplicated programming found on WOSU AM except for evening hours when it broadcast classical music.

1950: A popular daily program *In the Bookstall* premieres with Dr. Wilson Dumble reading popular books and poetry. Gene Gerrard took over after a few years and, until 1980, brought the characters alive with his many voice impersonations.

1952: The FCC orders the allocation of two hundred forty-two channels for educational use after the success of the Joint Committee on Educational Television (JCET) in Washington, DC, led by I. Keith Tyler of Ohio State.

1954: Construction begins on WOSU TV studios at North Star Road west of campus.

1955: Fred Calland is named music director of WOSU radio and would become one of the station's most popular on-air personalities. His dismissal in 1966 caused a community uproar.

1956 (February 20): WOSU-TV signs on at UHF channel 34, the twentieth educational television station in the country. Richard Hull takes over as director of radio and television, replacing Robert Higgy.

1957: WOSU-AM and FM originate the first live "stereophonic" local music program series. Both AM and FM frequencies are used, and listeners must have two radios to receive the stereo effect.

1958: With the acquisition of remote TV equipment and a used bus, WOSU TV begins broadcasting sports for the first time in February with the Ohio State–Michigan basketball game.

1958: Remedial Math would be the first of many courses produced and broadcast by WOSU-TV for students at Ohio State.

1959: WOSU-TV is awarded a grant from the Ford Foundation to buy Ohio's first videotape recorder, allowing the station to record programs for the first time.

1959: *On The Way Home,* a daily afternoon news program created by Don Quayle premieres on WOSU radio. Quayle later became the first president of NPR. The program would be one of the prototypes for NPR's *All Things Considered.*

1961: Twenty-one television courses are provided by WOSU on campus serving sixteen thousand students to assist the university as the student population surges. Courses are broadcast for Columbus Public Schools as part of the Greater Columbus Area ETV Project.

1967: President Lyndon Baines Johnson signs the Public Broadcasting Act.

1968: Ground-breaking ceremonies take place for the $5.7 million Center for Tomorrow (later named for President Novice Fawcett), which would house continuing studies, conference facilities, alumni affairs, and WOSU radio and television.

1968: WOSU-TV's first color broadcast is the Ohio State–Michigan football game in collaboration with ABC Sports. UHF antennas fly off the shelves of local electronics stores.

1969: A new national television network, the Public Broadcasting System (PBS), is incorporated. National Public Radio (NPR) is created a year later. The Corporation for Public Broadcasting is formed to support national programming and help fund local stations. *Sesame Street* premieres on PBS stations on November 10.

1970: After fifty years, WOSU radio moves from its WWI hangar studios to the new Fawcett Center for Tomorrow. TV would join over the next two years.

1970: The Ohio State campus is closed due to the campus riots, and the WOSU radio and television stations are ordered to go off the air for eight days.

1971: WOSU radio becomes a charter member of National Public Radio and on May 3 begins airing NPR's new afternoon news magazine, *All Things Considered.*

1973: University budget cuts force WOSU-TV to discontinue weekend broadcasts.

1973: The Friends of WOSU advisory and fund-raising board is formed. The board's first Friends-a-Thon pledge drive on TV and radio raises over $20,000.

1973: WOSU-FM begins broadcasting in stereo after its original 1949 transmitter is finally replaced. A 1,000-foot TV tower is built in Westerville, and the TV coverage radius is nearly doubled to 60 miles.

1974: WPBO-TV in Portsmouth signs on and is directed by the State of Ohio until its license is transferred to OSU in 1977.

1975: Using a subcarrier channel on 89.7 FM, the first radio reading service for the blind in Ohio is created. The Central Ohio Radio Reading Service would later change its name to VoiceCorps.

1978: Tom Ewing, a local teacher and bluegrass musician, convinces WOSU radio to air a weekly show called *Bluegrass Ramble.*

1979: After three years at the helm of WOSU, Elizabeth Young leaves and is replaced by Dale Ouzts.

1980: WOSU-AM becomes News 820, and WOSU-FM becomes Classical 89.7, with nineteen hours daily of classical music. For the first time, the stations totally separated their formats.

1980: *Airfare* magazine is created as a monthly viewer and listener guide for WOSU members.

1982: A new radio call-in program, *Open Line,* premieres with Lynn Neary as its first host. Neary and WOSU reporter Brian Naylor would soon move to NPR.

1983 (September 7): The *In The Know* high school quiz show moves to WOSU from WBNS and is hosted by Bill Schiffman. It would be the longest continuous television show in station history and was still being produced in 2018 with Schiffman as host.

1984: WOSU-TV starts an annual two-week television auction to raise funds for programming. The auction fund-raisers would continue for twenty-five years to supplement revenue.

1987: Compact discs, or CDs, start replacing vinyl records on WOSU-FM. The station holds a seminar on CD media for interested listeners.

1989: Fred Andrle takes over as host of *Open Line,* and the two-hour radio talk show would grow a devoted audience and define WOSU-AM in Columbus.

1989–98: WOSU signs on a series of FM stations to create a classical network: WOSV-FM Mansfield; WOSP-FM Portsmouth; WOSE-FM Coshocton; and WOSB-FM Marion.

1993: Descriptive Video Service (DVS) makes select public television programs on WOSU-TV accessible to people with visual impairments.

1998: WOSU unveils its first website, wosu.org.

2002: Thomas Rieland becomes general manager of WOSU after the retirement of Dale Ouzts.

2003: WOSU-TV goes digital and provides the first multichannel digital signal in Columbus.

2003: WOSU-TV begins a collaboration with the Greater Columbus Arts Council to provide coverage of the performing arts and local artists of all kinds in central Ohio. Programming would include *ArtZine* and *Broad & High*.

2004: WOSU-FM becomes the first radio station in central Ohio to broadcast in HD Radio digital technology.

2006: Under its new brand, WOSU Public Media opens a $5.6 million digital media center, including TV studios and production suites, at COSI in downtown Columbus. It would be the site of public forums, debates, and performances for the next ten years.

2009: WOSU is honored by the Corporation for Public Broadcasting for its creative efforts to help many poor and elderly viewers retain their television signal during the transition from analog to digital broadcasting.

2009: After the retirement of *Open Line* host Fred Andrle, WOSU premieres *All Sides with Ann Fisher* as WOSU radio's two-hour daily talk program.

2010: WOSU purchases commercial radio station 101.1 FM, known as CD101, and moves its classical music service to the station, branding it Classical 101.

2010: *Columbus Neighborhoods: Short North* debuts. It is the first of twelve popular historical/cultural documentaries about the community over the next six years.

2011: WOSU begins airing an all-day NPR News service on 89.7 FM in central Ohio and sells its historic 820 AM frequency to St. Gabriel Radio in Columbus.

2014: WOSU creates a model editorial integrity document signed by university leadership protecting the station from licensee interference in programming and news coverage.

2015: WOSU's profile of a man in drag who sells Tupperware is featured on the arts show *Broad & High* and receives over two million online views thanks to Facebook.

2016: After ten years, the WOSU studios at COSI are dismantled as COSI installs a dinosaur exhibit in partnership with the American Museum of Natural History. WOSU announces its intention to build a new headquarters in the 15th Avenue and High Street arts district development.

2017: WPBO-TV in Portsmouth is sold by Ohio State for $8.8 million as part of the FCC spectrum auction to bring more wireless broadband capacity to southern Ohio. On October 25, WPBO goes off the air after forty-three years.

2017: The educational technology agency Instructional Telecommunications Services of Central Ohio (ITSCO) merges into WOSU.

2018: An individual gift of $5 million, the largest in WOSU history, by Andy and Sandy Ross to support the new WOSU headquarters is made public.

2019: Groundbreaking to take place for the new WOSU Public Media headquarters at 14th Avenue and Pearl Alley.

WOSU STAFF AND BOARD

OCTOBER 1, 2018

Jay Kasey, Senior Vice President, The Ohio State University—Administration and Planning, serves as WOSU's administrative leader within the institution. The WOSU staff includes eight division leaders and a dozen other managers and supervisors.

WOSU DIVISION LEADERS

Thomas Rieland, General Manager
Mary Alice Akins, Business Office and Human Resources
Brent Davis, Arts, Life, and Culture
Mike Thompson, News & Public Affairs
Nick Houser, Digital Media
Mike Meadows, Technology
Marcelita Haskins, Educational Services
Amy Palermo, Educational Technology
Meredith Hart, Marketing & Communications

WOSU STAFF

Laura Baker
Ben Bays

Diana Bergemann
David Blankestyn
Steve Brown
Ellen Cahill
Dave Carwile
Greg Coulson
Elizabeth Curtis
Rob Dall
Michael De Bonis
Cheryl Dring
Joan Duffey
Nick Evans
Ann Fisher
Eric French
Cindy Gaillard
Alysia Gobert-Smith
Jennifer Hambrick
Stacia Hentz
Micki Hernandez
Ryan Hitchcock
Rhonda Holland
Debbie Holmes
Rick Hood
Amy Juravich
Tim Kelly
Connie Knoechel
Boyce Lancaster
George Levert
Dana Mack
Darrin McDonald
Kyle McKinnon
Victoria Miller
Bill Mooney
Adora Namigadde
Chuck Oney
Gary Orr
Casey Paris
Courtney Paul
Paul Peloquin

Kevin Petrilla
Paige Pfleger
Jan Pollard
John Prosek
Christopher Purdy
Kate Quickel
Mary Rathke
Jaclyn Reith
Gabe Rosenberg
Clare Roth
Christine Sadic
Richard Sanders
Hannah Schobel
Jackie Shafer
Susan Stonick
Kent Teeters
Pat Testa
Emily Thompson
Steve Thompson
Rob Walker
George Welling
Leticia Wiggins
Marylee Williams

WOSU HIGGY-TYLER STAFF AWARD, 1997–2018

The I. Keith Tyler Staff Award was combined with the Robert C. Higgy Staff Award in 2005. This list represents all honorees of either award, which is given annually by members of the WOSU Friends Advisory Board.

Mary Alice Akins
Joan Duffey
Herb Howenstine
Carole Browne
Cheryl Petrilla
John Prosek
Alysia Gobert-Smith

Paul Peloquin
Tom Lahr
Marcelita Haskins
Michael Thompson
Phyllis Madry
Ling Ling Sun
Fred Andrle
Mike Meadows
Edwin Clay
David Carwile, Carl Higginbotham, and Tom Lahr (Shared Award)
Sheri Walker
Courtney Paul
Beverley Ervine
Wendy Craver
Boyce Lancaster
Jan Pollard
Brent Davis
Diana Bergemann
Cindy Gaillard

BOARD PRESIDENTS OF FRIENDS OF WOSU, 1972–2018

Kathy McGinnis, 2018–
William Schiffman, 2016–18
Karen Ickes, 2014–16
Dale Heydlauff, 2012–14
Stephen Weed, 2009–12
Dan L. Cvetanovich, 2007–9
Rebecca P. Damsen, 2005–7
Laura Ecklar, 2003–5
David L. Baker, 2001–3
Kenneth R. Cookson, 1999–2001
Edmund T. King, 1997–99
Judy Ross, 1995–97
Don G. Davis, 1993–95
Michael A. Mess, 1991–93
Joseph L. Davis, 1989–91

Ruth W. Mount, 1987–89
Ken Keller, 1985–87
Jean W. Reilly, 1983–85
Spencer A. Schilling, 1981–83
Charles F. Glander, 1979–81
Ann Mall, 1977–79
William A. McNutt, 1975–77
Jane Cook, 1972–75

NOTES AND SOURCES

CHAPTER 1: THE FIRST SPARKS

Marconi's history is well documented, but I used *Thunderstruck* by Erik Larson (Broadway Books, 2007) as a key resource. Robert Marriott captured data from his 1900–1901 campus experiments in an undergraduate thesis paper, "Experiments in Wireless Telegraphy." His later campus visits were reported in the *Lantern*, and in the 1920s he wrote several articles for *Radio Broadcast* magazine and the Ohio State Monthly about his early wireless work at Ohio State and on Catalina Island.

CHAPTER 2: SOS COLUMBUS

Geoff Williams describes the disaster of the nation's most widespread flood in his authoritative book, *Washed Away* (Pegasus Books, 2014). Articles from the *Columbus Evening Dispatch* and *Columbus Citizen* outline Ackerberg's reporting from the Hilltop area, as does the *History of Ohio* by Charles Galbreath (American Historical Society, 1925). Akerberg is mentioned in a William S. Paley biography as a "dour, chain smoking, high-pressure salesman." A *Modern Electronics Magazine* article, "The Wireless Amateur in Times of Disaster," was written by B. N. Burglund for the April 1913 issue. Burglund was the wireless operator at the University of Michigan and honored James Mercer's dedication at the key on the Ohio State campus. *Radio's Hidden Voice* (University of Illinois Press, 2009) outlines some of the earliest history of educational radio before WWI, including the 1916 wireless network developed by the US Weather Bureau.

CHAPTER 3: WWI AND COLUMBUS RADIO

Randall Davidson has written one of most detailed profiles of an early educational radio station, *9XM Talking: WHA Radio and the Wisconsin Idea* (University of Wisconsin Press, 2006). The book includes multiple references to Ohio State's radio efforts and describes how the two stations were pioneers in the 1920s. I refer several times throughout this book to Wisconsin's broadcast efforts using *9XM Talking* as a reference. This chapter was enhanced by an interview with Joyce Higgy Miller, one of Bob Higgy's daughters. The descriptions of the WWI hangar and the installation of the radio station are from the *History of the Department of Electrical Engineering,* a 1970 departmental history written by Emerson Kimberly. On May 16, 1918, the *Lantern* published a short piece about a class in radio that had sent words across the Olentangy River to a radio receiver. It was the first mention of sending words instead of Morse code by wireless on campus. The early history of KDKA is captured in D. G. Little's paper, "KDKA—The radio broadcasting station of the Westinghouse Electric and Manufacturing Company at East Pittsburgh, Pennsylvania," published in June 1924. Lewis Chubb worked at Westinghouse for forty-two years and held over two hundred patents in science and engineering.

CHAPTER 4: A RADIO EXPLOSION

The *Ohio State Engineer* magazine told the story of the engineering fraternity pulling the wool over the ears of participants who came into their tents to hear KDKA. Eleanor Farjeon wrote the poem "F is for Forecast," which is part of a fascinating compilation called *Radio Waves: Poems Celebrating the Wireless* (Enitharmon Press, 2004). The *Lantern* and local newspapers provided information about the WEAO coverage of OSU football and Ohio State Day.

CHAPTER 5: PUTTING ON A SHOW

The Ohio State archives have the original signed pact that created the silent period by agreement among Columbus radio stations in 1922. A brief history of the silent nights is covered in *Selling Radio* by Susan Smulyan (Smithsonian Press, 1994).The issues surrounding the perceived quality of the female voice on air is described in some detail in *Radio Voices* (University of Minnesota Press, 1997). The reaction to hearing the voice of Ann Charles was documented in the university archives from letters to the station. A profile of WEAO written by Charles and Higgy was published on July 15, 1925 in *The Ohio State University Studies* magazine and served as background for this chapter. The efforts of Charles Wright to seek funding is covered in the station and OSU presidential papers in the OSU archives.

CHAPTER 6: FIGHTING FOR AIR

The landmark three-volume work (which concludes in the 1950s) by media historian Erik Barnouw on the history of broadcasting is used as a source at times throughout this book (Oxford University Press, 1966). *Radio's Hidden Voice* does a phenomenal job covering the creation and influence of the Federal Radio Commission and the challenges of early educational radio stations. The OSU *Lantern* online archive provided access to much of the programming information related to WEAO throughout the 1920s and '30s. It was an invaluable source of information.

CHAPTER 7: OHIO SCHOOL OF THE AIR

Benjamin Darrow wrote two books—*Radio, The Assistant Teacher* (R. G. Adams & Company, 1932) and *Radio Trailblazing* (College Book Company, 1940), which described the challenges and opportunities of education by radio. The "Ohio School of the Air" scrapbook in the OSU archives is full of newspaper clips and photographs. Michael Taylor wrote a dissertation on the life of Darrow, while the definitive book on early radio education is by William Bianchi, *Schools of the Air* (McFarland & Company, 2008). Dr. S. E. Frost covered OSU's groundbreaking radio research in his 1937 book, *Education's Own Stations* (University of Chicago Press, 1937). Research by Dr. Fred Lumley can be found in a series of monograms from the OSU Bureau of Educational Research archives. His mysterious death was covered in newspapers across the country. Lumley's father was a prominent OSU professor, who appeared often on WEAO.

CHAPTER 8: THRIVING IN THE 1930s

I developed the "Sad Sam" story after receiving a tip by historian and writer Dan McQuigg. The story is outlined in the book, *Sid Gillman: Father of the Passing Game* (Clerisy Press, 2012). S. E. Frost's *Education's Own Stations* (Arno Press, 1971) covered the difficulties many stations had surviving the 1930s. Two stories in *The Ohio State University Monthly* from March 1931 and March 1935 described station history and programming.

CHAPTER 9: WWII AND FM RADIO

The 1938 internal review of WOSU radio was found in the Rightmire papers. Alan Stavitsky's Ohio State dissertation, "From Pedagogy to Public: The Development of U.S. Public Radio's Audience Centered Strategies—WOSU, WHA, and WNYC, 1930–1987," was helpful in tracking the many objections the university made in Washington to seek better radio frequencies. The frequency chart is from his dissertation. A 1992 anniversary piece in *Airfare* about WEAO includes Margaret Tyler's story about the US invasion. This section taps into 2016 and 2017 interviews with WOSU veterans David Ayers, Bill McCormick, Thom McCain, and Robert Butche. The Bricker information comes from *Lantern* coverage, WOSU files, and the William Shkurti book, *The Ohio State University in the Sixties* (The Ohio State University Press, 2016). Harrison Summers presented his report about improving WOSU radio to the Radio Advisory Committee on February 20, 1949. As we move into the 1950s, Don Quayle was interviewed by Burt Harrison in 1978 as part of a series of oral histories with leaders of the early public broadcast movement. Ken Keller describes his discussions with Quayle about "On Your Way Home" in his entertaining book, *Monkey on a Turntable* (Orange Frazier Press, 2007). I. Keith Tyler wrote a piece as part of his research called "Radio as a Cultural Agency" (National Committee on Education by Radio, 1934), which described early efforts to set aside frequencies for educational radio. The story about Gene Gerrard showing up to interview James Thurber at his hotel room is from a personal interview with Don Davis.

CHAPTER 10: A CLASSICAL TRADITION

This chapter uses a portion of the information from a chapter in George Douglas's book *The Early Days of Radio Broadcasting* (McFarland & Co., 1987). The *Lantern* focused several stories on the Vivian record collection and the career of WOSU classical music personality Fred Calland. Thanks to John Schmidt and Chuck Pennington for providing the rare footage promoting stereo radio on WOSU TV, and to Donald Lakin for telling his personal stereo story in response to one of my history blogs on the WOSU website.

CHAPTER 11: SAVING EDUCATIONAL TELEVISION

Richard Hull describes the creation of the JCET in "A Note on the History Behind ETV," in *Educational Television: The Next Ten Years* (The Institute for Communications Research, 1962). The activities of the Joint Council and FCC hearings are covered extensively in the I. Keith Tyler Papers in The Ohio State University archives and in *Broadcasting Magazine.* Tyler provided an informative oral history to Burt Harrison as part of the Public Radio Oral History Project.

CHAPTER 12: UHF 34

The Howard Bevis and I. Keith Tyler papers in the OSU archives were prominent sources for this chapter. Also valuable were personal interviews with Robert Butche, John Schmidt, and David Ayers, and the unpublished memoir of WOSU engineer Lee Howard. Schmidt's personal film collection, which is extensive, included the kinescope of his short promotion for the WOSU TV station. A WOSU booster, Michael Schmidt, recalled for me his family's experience with early WOSU TV. Kenneth Bartter's 1961 OSU thesis, "An Historical Investigation Illustrating the Evolution of Programming of WOSU-TV from 1956 through 1960," was a helpful source. The Airborne Television project was unique and covered widely as the latest in educational innovation. The examination of faculty and student attitudes toward television in the OSU classrooms is well documented in a 1967 OSU thesis called "Attitudes toward Television" by William Buccalo. An interview with John Morison, who supervised the telecourse programming and provided other information about the statewide network was very beneficial, as were the materials and information provided by Gene Gilliom.

CHAPTER 13: SPARKS FLEW

The interview with A. V. Shirk, who worked alongside Fred Calland, revealed more behind-the-scenes fun at the radio operations than I could share. Personal interviews with Mary Hoffman and Charlie Pickard were also helpful. Don Quayle recalled his four years at WOSU for Burt Harrison in an interview that was part of the Public Radio Oral History Project. The audio party reel created by WOSU radio staff is in WOSU's archives. Richard Hull's memos are preserved in his papers at Ohio State. The Fawcett Papers include dozens of letters protesting the Fred Calland firing. Ken Keller's book was again used to quote the story about Fred Calland stripping in the studio. After his death in 1999, Fred Calland's sister contributed his massive sound recordings collection, which had ballooned to a remarkable forty thousand, to the University of Missouri Special Collections in Kansas City. Calland had many talents, including that of a quilt designer and quilt maker.

CHAPTER 14: CAMPUS CHAOS

The 1967 Public Broadcasting Act and the LBJ speech are covered in the *Current Magazine* series on the history of public media and Jack Mitchell's book, *Listener Supported.* Bill Shkurti's book about the 1960s on campus was a key source about the issues surrounding the chaos at Ohio State in 1969 and 1970. The WOSU files helped outline WOSU's coverage of the riots, but an interview with Don Davis and John Mount, who was the vice-president assigned to supervise WOSU in 1970, also contributed. The lead-up to the OSU–Michigan game and WOSU's role received wide coverage, including a piece in the *Chicago Tribune* (November 22, 1968). David Jensen's 1970 thesis about the history of WOSU radio, "An Historical Descriptive Analysis of The Ohio State University Radio Stations WOSU and WOSU FM from 1956 through 1970," outlines the CPB ombudsman grant and the challenging environment within WOSU radio during the late 1960s. Anne Gudenkauf provided information about the ombudsman project and her time at WOSU during a personal interview.

CHAPTER 15: TIME AND CHANGE

The Hull and Fawcett Papers provided ample information about the plans for the Fawcett Center for Tomorrow. Other sources include the *Lantern* and an oral history with Robert Holsinger. The *Black Studies Journal* information is sourced from WOSU files. The building at the statewide broadcast agency, currently called the Broadcast Educational Media Commission, is named in Richard Hull's honor. It is the same building where WOSU-TV began operations in 1956 on North Star Road. I appreciate the assistance of Brent Davis on the *Bluegrass Ramble* section.

CHAPTER 16: GROWING PAINS

Interviews with Dale Ouzts, John Prosek, Lynn Neary, and Fred Andrle helped develop this section. My access to financial information going back to the 1950s was helpful in tracking OSU support and membership growth. In addition, WOSU published annual reports starting in 1980 that reviewed fund-raising and programming. Alan Stavitsky's 1990 dissertation provided information about the development of the Friends Board and the ongoing funding challenges the station faced. I appreciate Joyce Higgy Miller's information about her father for the piece about his last years. Dr. Thom McCain provided valuable perspective on the lengthy university review of WOSU. The report about the WOSU review was leaked to *Columbus Monthly* in November 1988, sparking coverage in the *Lantern* and the *Cleveland Plain Dealer.* The description of the impact of *Tongues Untied* on public stations across the country is from *A History of Public Broadcasting* and local coverage.

CHAPTER 17: BEYOND OSU

Personal interviews with Bill Schiffman, Mary Hoffman, and Howard Ornstein helped provide important perspective for this section. Bill

VanDervoort's daughter recalled his love of the Met and WOSU during his memorial service. Near the end of his life, Bill would regularly call me to vociferously praise WOSU programming. The report by Dean Joan Huber, published March 1, 1996, was titled "On the WOSU Stations' Relation to Ohio State: Some Comparisons." The student radio section was developed predominantly from *Lantern* articles going back to 1940.

CHAPTER 18: PERSPECTIVE: 2003–2018

Much of this chapter is informed by WOSU files and annual reports over the past fifteen years and by my own personal history at WOSU Public Media during this period.

BIBLIOGRAPHY

Adams, Mike. *Columbus Radio*. Charleston: Arcadia Publishing, 2016

Balas, Glenda R. *Recovering a Public Vision for Public Television*. Lanham, MD: Rowman & Littlefield Publishers, 2003.

Barnouw, Eric. *A Tower in Babel: A History of Broadcasting in the United States to 1933*. New York: Oxford University Press, 1966.

Bianchi, William. *Schools of the Air.* Jefferson, NC: McFarland & Company, 2008.

Brinson, Susan L. *Personal and Public Interests: Frieda B. Hennock and the Federal Communications Commission*. Westport, CT: Praeger Publishing, 2002.

Bryson, Bill. *One Summer: America 1927*. New York: Doubleday, 2013.

Coe, Lewis. *Wireless Radio: A Brief History.* Jefferson, NC: McFarland & Company, 1996.

Darrow, Ben. *Radio, The Assistant Teacher.* Columbus: R. G. Adams & Company, 1932.

Darrow, Ben. *Radio Trailblazing*. Columbus: College Book Company, 1940.

Davidson, Randall. *9XM Talking: WHA Radio and the Wisconsin Idea*. Madison: University of Wisconsin Press, 2006.

Desoto, Clinton. *200 Meters & Down*. West Hartford, CT: The American Radio Relay League, Inc., 1936.

Douglas, George H. *The Early Days of Radio Broadcasting.* Jefferson, NC: McFarland & Co., 1987.

Douglas, Susan. *Listening In: Radio and the American Imagination.* Minneapolis: University of Minnesota Press, 2004.

Engelman, Ralph. *Public Radio and Television in America: A Political History.* Thousand Oaks, CA: SAGE Publications, 1996.

Frost, S. E. *Education's Own Stations.* Chicago: University of Chicago Press, 1937.

Harrison, Burt. *Washington State: On the Air.* Pullman, WA: KWSU Radio-Television Services, 1993.

Hilmes, Michelle. *Radio Voices—American Broadcasting, 1922–52.* Minneapolis: University of Minnesota Press, 1997.

Hinds, Conrade C. *Columbus and the Great Flood of 1913.* Charleston, SC: The History Press, 2013.

Galbreath, Charles. *History of Ohio.* Chicago: American Historical Society, 1925.

Goerler, Raimund. *The Ohio State University: An Illustrated History.* Columbus: The Ohio State University Press, 2011.

Gregory, Danny and Paul Sahre. *Hello World—A Life in Ham Radio.* New York: Princeton Architectural Press, 2003.

Hawes, William. *Public Television: America's First Station.* Santa Fe, NM: Sunstone Press, 1996.

Hill, Ruane and Allen Koenig, eds. *The Farther Vision: Educational Television Today.* Madison: University of Wisconsin Press, 1967.

Katzowitz, Josh. *Sid Gillman: Father of the Passing Game.* Clerisy Press, 2012.

Keller, Ken. *Monkey on a Turntable: Living on Love and Leftovers in Local Radio.* Wilmington, OH: Orange Frazier Press, 2007.

Kinnison, William. *Building Sullivant's Pyramid: Administrative History of The Ohio State University, 1870–1907.* Columbus: The Ohio State University Press, 1970.

Larson, Eric. *Thunderstruck.* New York: Broadway Books, 2007.

Lentz, Ed. *Columbus: The Story of a City.* Charleston, SC: Arcadia Publishing, 2003.

Lewis, Tom. *Empire of the Air: The Men Who Made Radio.* New York: HarperCollins, 1993.

Lundy, Bert. *Telegraph, Telephone & Wireless: How Telecom Changed the World*. Bert Lundy, 2008.

Mitchell, Jack. *Listener Supported: The Culture and History of Public Radio*. Westport, CT: Praeger Publishers, 2005.

Mitchell, Jack. *Wisconsin on the Air.* Madison: Wisconsin Historical Society Press, 2016.

Razlogova, Elena. *Early Radio and the American Public.* Philadelphia: University of Pennsylvania Press, 2011.

Robertson, Jim. *Televisionaries.* Charlotte Harbor, FL: Tabby House Books, 1993

Rudel, Anthony. *Hello Everybody! The Dawn of American Radio.* San Diego: Harcourt, 2008.

Shkurti, William. *The Ohio State University in the Sixties: The Unraveling of the Old Order.* Columbus: The Ohio State University Press, 2016.

Siepmann, Charles A. *Radio's Second Chance.* Boston: Little, Brown and Company, 1946.

Slotten, Hugh Richard. *Radio's Hidden Voice.* Urbana and Chicago: University of Illinois Press, 2009.

Smulyan, Susan. *Selling Radio.* Washington and London: Smithsonian Press, 1994.

Street, Sean. *Radio Waves: Poems Celebrating the Wireless.* London: Enitharmon Press, 2004.

Wagner, Paul. *Radio Journalism.* Minneapolis: Burgess Publishing, 1940.

Weaver, David. *Black Diva of the Thirties: The Life of Ruby Elzy.* Jackson: University Press of Mississippi, 2004.

Williams, Geoff. *Washed Away.* New York: Pegasus Books, 2014.

Witherspoon, John, Roselle Kovitz, Robert Avery, and Alan Stavitsky. *A History of Public Broadcasting. Current, Newspaper for People in Public Media.* Washington, DC, 2000.

OTHER NOTABLE REFERENCES

Airfare magazine, WOSU's monthly program and information guide (1980–present).

Bartter, Kenneth L. "An Historical Investigation Illustrating the Evolution of Programming of WOSU-TV from 1956 through 1960." Master's Thesis. The Ohio State University, 1961.

Carnegie Commission on Educational Television. "Public Television: A Program for Action. Report and Recommendations." New York: Harper & Row, 1967.

Harrison, Burt. Interviews conducted for the Public Radio Oral History Project funded by the Corporation for Public Broadcasting. Interviews with I. Keith Tyler, Donald Quayle, Richard Hull, and Elizabeth Young, 1978.

Howard, Edgar L. "The Telcom Story—An Ongoing Recollection." Unpublished paper provided to the author, n.d.

Hull, Richard B. "Educational Television in the United State—Status Report 1957." Report prepared for the Educational Television & Radio Center, the Fund for Adult Education, and The Ohio State University, 1957.

Jensen, David R. "An Historical Descriptive Analysis of The Ohio State University Radio Stations WOSU and WOSU FM from 1956 through 1970." Master's Thesis. The Ohio State University, 1970.

Kimberly, Emerson. "History of the Department of Electrical Engineering." *The Ohio State University Monthly*, 1970.

Knepper, William E. "This Is WEAO Broadcasting." *The Ohio State University Monthly*, 1931.

Madden, Richard. "A History of WOSU Radio 1922–1956." Master's Thesis. The Ohio State University, 1968.

Marriott, Robert. "Experiments in Wireless Telegraphy." Master's Thesis. The Ohio State University, 1901.

Olsen, O. Joe, ed. *Education of the Air.* Twenty-Third Yearbook of the Institute for Education by Radio and Television. The Ohio State University, Columbus, 1953.

Schmid, William T. "A Historical Analysis of the Educators' Request for Non-Commercial Television Channel Reservations in the United States." PhD Dissertation. The Ohio State University, 1970.

"Station WOSU—The Voice of Ohio State University." *Ohio State Engineer Magazine,* February 1936.

Stavitsky, Alan. "From Pedagogy to Public: The Development of U.S. Public Radio's Audience Centered Strategies—WOSU, WHA, and WNYC, 1930–1987." PhD Dissertation. The Ohio State University, 1990.

Taylor, Michael. "Ben Darrow and The Ohio School of the Air." PhD Dissertation. The Ohio State University, 1974.

"This Is WEAO Broadcasting." *Ohio State University Monthly*, March 1931.

Tyler, I. Keith. The I. Keith Tyler Papers, The Ohio State University archives.

Weisz, Henry. "820 on Your Dial." *Ohio State Engineer Magazine,* 1944.

"WOSU External Program Evaluation." The Ohio State University, May, 1988.

"WOSU Internal Program Review." The Ohio State University, September, 1987.

"WOSU—A Study of Its Organization." Report prepared for Vice President J. L. Morrill, April 1939.

A fun staff photograph at the 2017 holiday party by
Friends Board member and professional photographer Ray LaVoie.